T0235588

Narrowband Direction of Arrival Estimation for Antenna Arrays

© Springer Nature Switzerland AG 2022
Reprint of original edition © Morgan & Claypool 2008

All rights reserved. No part of this publication may be reproduced, stored in a retrieval system, or transmitted in any form or by any means—electronic, mechanical, photocopy, recording, or any other except for brief quotations in printed reviews, without the prior permission of the publisher.

Narrowband Direction of Arrival Estimation for Antenna Arrays
Jeffrey Foutz, Andreas Spanias, and Mahesh K. Banavar

ISBN: 978-3-031-00409-4   paperback

ISBN: 978-3-031-01537-3   ebook

DOI: 10.1007/978-3-031-01537-3

A Publication in the Springer series

SYNTHESIS LECTURES ON ANTENNAS #8

Lecture #8

Series Editor: Constantine A. Balanis, Arizona State University

Series ISSN

ISSN 1932-6076   print
ISSN 1932-6084   electronic

Narrowband Direction of Arrival Estimation for Antenna Arrays

Jeffrey Foutz, Andreas Spanias, and Mahesh K. Banavar
Arizona State University

SYNTHESIS LECTURES ON ANTENNAS #8

ABSTRACT

This book provides an introduction to narrowband array signal processing, classical and subspace-based direction of arrival (DOA) estimation with an extensive discussion on adaptive direction of arrival algorithms. The book begins with a presentation of the basic theory, equations, and data models of narrowband arrays. It then discusses basic beamforming methods and describes how they relate to DOA estimation. Several of the most common classical and subspace-based direction of arrival methods are discussed. The book concludes with an introduction to subspace tracking and shows how subspace tracking algorithms can be used to form an adaptive DOA estimator. Simulation software and additional bibliography are given at the end of the book.

KEYWORDS

smart antennas, array processing, adaptive antennas, direction of arrival, DSP

Contents

CHAPTER 1

Introduction

Propagating fields are often measured by an array of sensors. A sensor array consists of a number of transducers or sensors arranged in a particular configuration. Each transducer converts a mechanical vibration or an electromagnetic wave into a voltage. Acoustic waves occur in microphone or sonar array applications. Mechanical waves are associated with seismic exploration and electromagnetic waves are used in wireless communications. Array signal processing applications include radar, sonar, seismic event prediction, microphone sensors, and wireless communication systems [1].

In engineering applications, where an incoming wave is detected and/or measured by an array, the associated signals at different points in space can be processed to extract various types of information including their direction of arrival (DOA). Algorithms for estimating the DOA in antenna arrays are often used in wireless communications to increase the capacity and throughput of a network. In this book, the focus will be on antenna arrays that receive or transmit electromagnetic

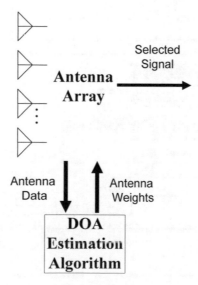

FIGURE 1.1: Antenna array and direction of arrival algorithms.

waves in a digital communication network. Although most of the algorithms presented will focus on radio frequencies, we note that many of the discussed concepts can also be applied to mechanical and acoustic waves. We also note that the array processing algorithms presented can be used for real-time or offline applications.

DOA methods can be used to design and adapt the directivity of array antennas as shown in Figure 1.1. For example, an antenna array can be designed to detect the number of incoming signals and accept signals from certain directions only, while rejecting signals that are declared as interference. This spatiotemporal estimation and filtering capability can be exploited for multiplexing co-channel users and rejecting harmful co-channel interference that may occur because of jamming or multipath effects (Figure 1.2).

DOA algorithms can be divided into three basic categories, namely, classical, subspace methods, and maximum likelihood (ML) techniques [15]. In this book, the most important methods in each of these three categories will be discussed. The ML method offers high performance but

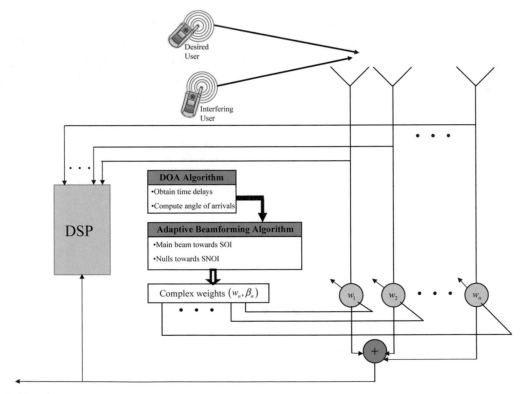

FIGURE 1.2: Antenna array used to spatially filter interference.

is computationally expensive. The subspace methods also perform well and have several computationally efficient variants. The classical methods are conceptually simple but offer modest or poor performance while requiring a relatively large number of computations. Note that these algorithms are initially presented under the assumption that the signal sources are stationary in space and that the incoming signals are not correlated (no signals present due to multipath propagation). At the end of the book, adaptive DOA estimation is discussed for the case where the directions of arrival are changing with time.

Classical methods covered in this book include the delay-and-sum method and the Minimum Variance Distortionless Response (MVDR) method. The subspace methods described include different versions of the Multiple Signal Classification (MUSIC) algorithm and the Estimation of Signal Parameters via Rotational Invariance Technique (ESPRIT). Among the ML techniques, the focus will be on the alternating projection algorithm (APA). The algorithms can operate with a variety of array geometries including uniform linear array (Figure 1.3) and the uniform planar array (Figure 1.4). Most or all the algorithms presented work with the uniform linear array. For the

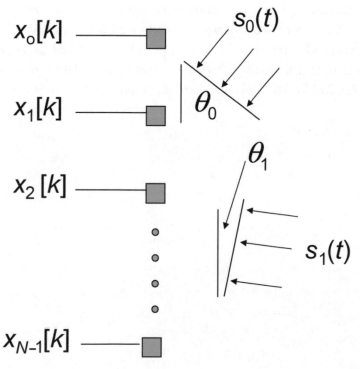

FIGURE 1.3: A uniform linear array.

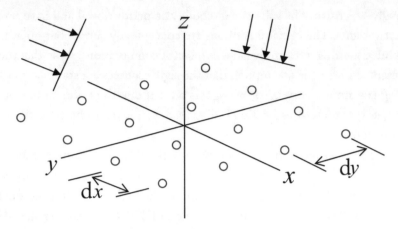

FIGURE 1.4: Illustration of a 4 × 4 uniform planar array.

uniform planar array, one can use the MUSIC or the 2D ESPRIT algorithms, whereas for semi-spherical arrays the MUSIC algorithm may be appropriate.

The book organization is as follows. In the second chapter, we provide background information on propagation delays of an electromagnetic signal across an array; the signal model is established based on a narrowband digitally modulated signal. The covariance matrix and its eigenstructure are described in detail. In Chapter 3, we describe classical DOA methods and high-resolution subspace methods. Finally, Chapter 4 covers adaptive DOA estimation and provides simulation examples. MATLAB software realizations of some of the algorithms are provided in the Appendix.

· · · ·

CHAPTER 2

Background on Array Processing

2.1 INTRODUCTION

This chapter presents the signal model for narrowband arrays. The structure of propagation delays is first discussed for a linear array geometry. The spatial covariance matrix is formed and its spectral decomposition is analyzed. Subspaces are formed by considering associations of eigenvalues and eigenvectors with the signal and noise components of the signal. This data model will be used throughout the book especially in explaining high-resolution direction of arrival (DOA) methods.

2.1.1 Propagation Delays in Uniform Linear Arrays

Consider a uniform linear array geometry with N elements numbered $0, 1, \ldots, N-1$. Consider that the array elements have half-a-wavelength spacing between them. Because the array elements are closely spaced, we can assume that the signals received by the different elements are correlated. A propagating wave carries a baseband signal, $s(t)$, that is received by each array element, but at a different time instant. It is assumed that the phase of the baseband signal, $s(t)$, received at element 0 is zero. The phase of $s(t)$ received at each of the other elements will be measured with respect to the phase of the signal received at the 0th element. To measure the phase difference, it is necessary to measure the difference in the time the signal $s(t)$ arrives at element 0 and the time it arrives at element k. By examining the geometry from Figure 2.1, and using basic trigonometry and facts from wave propagation, the time delay of arrival can be computed as:

$$\Delta t_k = \frac{kD \sin\theta}{c}, \tag{2.1}$$

where c is the speed of light.

Suppose $s(t)$ is a narrowband digitally modulated signal with lowpass equivalent $s_l(t)$, carrier frequency f_c, and symbol period T. It can be written as

$$s(t) = \operatorname{Re}\left\{s_l(t)e^{j2\pi f_c t}\right\}. \tag{2.2}$$

The signal received by the kth element is given by

$$x_k(t) = \operatorname{Re}\left\{s_l(t-\Delta t_k)\, e^{j2\pi f_c(t-\Delta t_k)}\right\}. \tag{2.3}$$

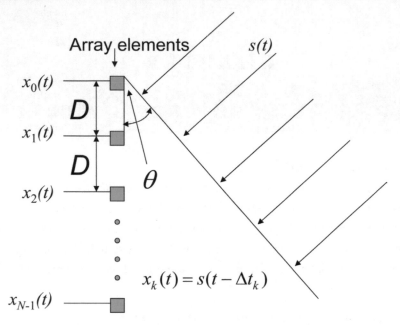

FIGURE 2.1: The propagating wave carries the signal $s(t)$ that is received by each element in the array but at a different time instant. Δt_k is difference in time of arrival of the signal at element 0 and element k, in seconds. c is the speed of the wave in m/s. D is distance between elements in meters.

Now suppose that the received signal at the kth element is downconverted to the baseband. In that case, the baseband received signal is:

$$x_k(t) = s_l(t - \Delta t_k)e^{-j2\pi f_c \Delta t_k}. \tag{2.4}$$

2.1.2 Narrowband Approximation

The received baseband signal is sampled with sampling period T seconds, which is also the symbol period, i.e.,

$$x_k(nT) = s_l(nT - \Delta t_k)e^{-j2\pi f_c \Delta t_k}. \tag{2.5}$$

In a wireless digital communication system, the symbol period will be much greater than each of the propagation delays across the array, that is,

$$T \gg \Delta t_k, \quad k = 0, 1, \ldots, N-1. \tag{2.6}$$

This allows the following approximation to be made [8].

$$x_k(nT) \approx s_l(nT)e^{-j2\pi f_c \Delta t_k}.$$ (2.7)

The constants c and f_c can be related through the equation $c = \lambda f_c$, where λ is the wavelength of the propagating wave. The element spacing can be computed in wavelengths as $d = D/\lambda$. Using these equations, (2.7) can be written as:

$$x_k(nT) \approx s_l(nT)e^{-j2\pi nd \sin\theta}.$$ (2.8)

To avoid aliasing in space, the distance between elements, D, must be $\lambda/2$ or less [7]. In the simulations shown in this book, we use $D = \lambda/2$ or $d = 1/2$, which simplifies (2.8) to:

$$x_k(nT) \approx s_l(nT)e^{-j\pi k \sin\theta}.$$ (2.9)

A discrete time notation will now be used with time index n so that (2.9) can be written as:

$$x_k[n] \approx s[n]e^{-j\pi k \sin\theta} = s[n]a_k(\theta).$$ (2.10)

Let the nth sample of the baseband signal at the kth element be denoted as $x_k[n]$. When there are r signals present, the nth symbol of the ith signal will be denoted $s_i[n]$ for $i = 0, 1, ..., r - 1$. The baseband, sampled signal at the kth element can be expressed as

$$x_k[n] \approx \sum_{i=0}^{r-1} s_i[n]a(\theta_i).$$ (2.11)

If the propagating signal is not digitally modulated and is narrowband, the approximation shown in (2.8) is still valid.

2.1.3 Matrix Equation for Array Data

By considering all the array elements, i.e., $k = 0, 1, 2, ..., N - 1$, equation (2.11) can be written in a matrix form as follows:

$$
\begin{bmatrix}
x_0[n] \\
x_1[n] \\
\cdot \\
\cdot \\
\cdot \\
x_{N-1}[n]
\end{bmatrix}
=
\begin{bmatrix}
a_0(\theta_0) & a_0(\theta_1) & .. & a_o(\theta_{r-1}) \\
a_1(\theta_0) & & \cdot & \cdot \\
\cdot & & \cdot & \\
\cdot & & \cdot & \cdot \\
\cdot & \cdot & \cdot & \\
a_{N-1}(\theta_0) & \cdot & .. & a_{N-1}(\theta_{r-1})
\end{bmatrix}
\begin{bmatrix}
s_0[n] \\
s_1[n] \\
\cdot \\
\cdot \\
s_{r-1}[n]
\end{bmatrix}
+
\begin{bmatrix}
v_0[n] \\
v_1[n] \\
\cdot \\
\cdot \\
v_{N-1}[n]
\end{bmatrix},
$$ (2.12)

where additive noise, $v_k[n]$, is considered at each element. The $N \times 1$ vector $\mathbf{x}_n$, the $N \times r$ matrix $\mathbf{A}$ along with the signal and noise vectors $\mathbf{s}_n$ and $\mathbf{v}_n$, respectively, can be used to write equation (2.12) in compact matrix notation, as follows:

$$\mathbf{x}_n = \begin{bmatrix} \mathbf{a}(\theta_0) \ \mathbf{a}(\theta_1) \ \dots \ \mathbf{a}(\theta_{r-1}) \end{bmatrix} \mathbf{s}_n + \mathbf{v}_n = \mathbf{A}\mathbf{s}_n + \mathbf{v}_n \ . \tag{2.13}$$

The columns of the matrix $\mathbf{A}$, denoted by $\mathbf{a}(\theta_i)$, are called the steering vectors of the signals $s_i(t)$. These form a linearly independent set assuming the angle of arrival of each of the r signals is different. The vector $\mathbf{v}_n$ represents the uncorrelated noise present at each antenna element. Because the steering vectors are a function of the angles of arrival of the signals, the angles can then be computed if the steering vectors are known or if a basis for the subspace spanned by these vectors is known [9].

The set of all possible steering vectors is known as the *array manifold* [9]. For certain array configurations, such as the linear, planar, or circular, the array manifold can be computed analytically. However, for other more complex antenna array geometries the manifold is typically measured experimentally. In the absence of noise, the signal received by each element of the array can be written as:

$$\mathbf{x}_n = \mathbf{A}\mathbf{s}_n. \tag{2.14}$$

It can be seen that the data vector, $\mathbf{x}_n$, is a linear combination of the columns of $\mathbf{A}$. These elements span the *signal subspace*. In the absence of noise, one can obtain observations of several vectors $\mathbf{x}_n$ and once r linearly independent vectors have been estimated, a basis for the signal subspace can be calculated. The idea of a signal subspace is used in many applications such as DOA [11], frequency estimation [10], and low-rank filtering [5].

2.1.4   Eigenstructure of the Spatial Covariance Matrix

The spatial covariance matrix of the antenna array can be computed as follows. Assume that $\mathbf{s}_n$ and $\mathbf{v}_n$ are uncorrelated and $\mathbf{v}_n$ is a vector of Gaussian, white noise samples with zero mean and correlation matrix $\sigma^2\mathbf{I}$. Define $\mathbf{R}_{ss} = E[\mathbf{s}_n\mathbf{s}_n^H]$. The spatial covariance matrix can then be written as

$$\mathbf{R}_{xx} = E\begin{bmatrix} \mathbf{x}_n\mathbf{x}_n^H \end{bmatrix} = E\begin{bmatrix} (\mathbf{A}\mathbf{s}_n + \mathbf{v}_n)(\mathbf{A}\mathbf{s}_n + \mathbf{v}_n)^H \end{bmatrix} = \mathbf{A}E\begin{bmatrix} \mathbf{s}_n\mathbf{s}_n^H \end{bmatrix}\mathbf{A}^H + E\begin{bmatrix} \mathbf{v}_n\mathbf{v}_n^H \end{bmatrix}$$

$$= \mathbf{A}\mathbf{R}_{ss}\mathbf{A}^H + \sigma^2\mathbf{I}_{N \times N}. \tag{2.15}$$

Since the matrix $\mathbf{R}_{xx}$ is Hermitian (complex conjugate transpose), it can be unitarily decomposed and has real eigenvalues. Now, let us examine the eigenvectors of $\mathbf{R}_{xx}$ and assume that N has been chosen large enough so that $N > r$. Any vector, $\mathbf{q}_n$, which is orthogonal to the columns of $\mathbf{A}$, is also an eigenvector of $\mathbf{R}_{xx}$, which can be shown by the following equation:

$$\mathbf{R}_{xx}\mathbf{q}_n = \left(\mathbf{A}\mathbf{R}_{ss}\mathbf{A}^H + \sigma^2\mathbf{I}\right)\mathbf{q}_n = 0 + \sigma^2\mathbf{I}\mathbf{q}_n = \sigma^2\mathbf{q}_n. \tag{2.16}$$

The corresponding eigenvalue of $\mathbf{q}_n$ is equal to σ^2. Because $\mathbf{A}$ has dimension $N \times r$, there will be $N - r$ such linearly independent vectors whose eigenvalues are equal to σ^2. The space spanned by these $N - r$ eigenvectors is called the *noise subspace*. If $\mathbf{q}_s$ is an eigenvector of $\mathbf{A}\mathbf{R}_{ss}\mathbf{A}$ then,

$$\mathbf{R}_{xx}\mathbf{q}_s = \left(\mathbf{A}\mathbf{R}_{ss}\mathbf{A}^H + \sigma^2\mathbf{I}\right)\mathbf{q}_s = \sigma_s^2\mathbf{q}_s + \sigma^2\mathbf{I}\mathbf{q}_s = \left(\sigma_s^2 + \sigma^2\right)\mathbf{q}_s \tag{2.17}$$

[7, 8, 15]. Note that $\mathbf{q}_s$ is also an eigenvector of $\mathbf{R}_{xx}$, with eigenvalue $(\sigma_s^2 + \sigma^2)$, where σ_s^2 is the eigenvalue of $\mathbf{A}\mathbf{R}_{ss}\mathbf{A}$. Since the vector $\mathbf{A}\mathbf{R}_{ss}\mathbf{A}\mathbf{q}_s$ is a linear combination of the columns of $\mathbf{A}$, the eigenvector $\mathbf{q}_s$ lies in the columnspace of $\mathbf{A}$. There are r such linearly independent eigenvectors of $\mathbf{R}_{xx}$. Again, the space spanned by these r vectors is the signal subspace. Note that the signal and noise subspaces are orthogonal to one another. Also, if the eigenvalues of $\mathbf{R}_{xx}$ are listed in descending order $\sigma_1^2, \ldots, \sigma_r^2, \sigma_{r+1}^2, \sigma_N^2$, then $\sigma_i^2 \geq \sigma_{i+1}^2$ for $i = 1, 2, \ldots, r - 1$ and $\sigma_r^2 > \sigma_{r+1}^2 = \sigma_{r+2}^2 = \cdots = \sigma_N^2 = \sigma^2$.

The eigendecomposition of R_{xx} can then be written as

$$\mathbf{R}_{xx} = \mathbf{Q}\mathbf{D}\mathbf{Q}^H = \begin{bmatrix} \mathbf{Q}_s & \mathbf{Q}_n \end{bmatrix} \begin{bmatrix} \mathbf{D}_s & 0 \\ 0 & \sigma^2\mathbf{I} \end{bmatrix} \begin{bmatrix} \mathbf{Q}_s & \mathbf{Q}_n \end{bmatrix}^H. \tag{2.18}$$

The matrix $\mathbf{Q}$ is partitioned into an $N \times r$ matrix $\mathbf{Q}_s$ whose columns are the r eigenvectors corresponding to the signal subspace, and an $N \times (N - r)$ matrix $\mathbf{Q}_n$ whose columns correspond to the "noise" eigenvectors. The matrix $\mathbf{D}$ is a diagonal matrix whose diagonal elements are the eigenvalues of $\mathbf{R}_{xx}$ and is partitioned into an $r \times r$ diagonal matrix $\mathbf{D}_s$ whose diagonal elements are the "signal" eigenvalues and an $(N - r) \times (N - r)$ scaled identity matrix $\sigma^2\mathbf{I}_{N \times N}$ whose diagonal elements are the $N \times r$ "noise" eigenvalues.

An alternative to finding the eigenvectors of the autocorrelation matrix is to use the data matrix $\mathbf{X}$. The rows of the matrix $\mathbf{X}$ are complex conjugate transpose of the data vectors obtained from the array of sensors. Suppose that the data matrix $\mathbf{X}$ contains K snapshots of data obtained from N sensors in a linear array. The matrix $\mathbf{X}$ is $K \times N$ and can be written as the product of three matrices:

$$\mathbf{X} = \mathbf{U}\mathbf{D}\mathbf{V}^H. \tag{2.19}$$

The matrix $\mathbf{U}$ is a $K \times K$ matrix whose columns are orthonormal, $\mathbf{D}$ is a diagonal $K \times N$ matrix, and $\mathbf{V}$ is an $N \times N$ matrix whose columns are also orthonormal. This decomposition is known as the singular value decomposition (SVD). The SVD of $\mathbf{X}$ is related to the spectral decomposition (eigendecomposition) of the spatial covariance matrix $\mathbf{R}_{xx}$. The columns of the matrix $\mathbf{V}$ will be eigenvectors of $\mathbf{R}_{xx}$ and the diagonal elements of the matrix $\mathbf{D}$ will be square roots of the eigenvalues of $\mathbf{R}_{xx}$. Because stable SVD algorithms are available, several methods rely on decomposing the data matrix instead of diagonalizing the spatial covariance matrix to obtain the basis for the signal subspace. In practice, the $N - r$ smallest eigenvalues will not be precisely σ^2; rather, they will all have small values compared to the signal eigenvalues. This is because the matrix $\mathbf{R}_{xx}$ is not known perfectly, but must be estimated from the data [12]. A common estimator for the spatial covariance matrix is the sample spatial covariance matrix, which is obtained by averaging rank-one data matrices of the form $(\mathbf{x}_n \mathbf{x}_n^H)$, i.e.:

$$\mathbf{R}_{xx} = \frac{1}{K} \sum_{i=0}^{K-1} \mathbf{x}_n \mathbf{x}_n^H, \tag{2.20}$$

where K is the total number of snapshots of data available from the sensors.

Although the discussion so far has focused on the uniform linear array, the principles of signal and noise subspaces also apply to other array geometries such as the uniform planar and the semispherical arrays.

2.2   ANTENNA BEAMFORMING BASICS

In this section, a brief introduction to beamforming is provided. Classical direction of arrival estimation algorithms are based on beamforming concepts; therefore, this discussion is included to give the reader the necessary background to understand Sections 3.1.1 and 3.1.2.

If a weighted linear combination of the output of each element is taken, the array output can be computed by

$$y[n] = \sum_{k=0}^{N-1} w_k x_k[n], \tag{2.21}$$

where w_k are the complex weights of the beamformer, which are shown in Figure 2.2. In vector notation, the output can be written as:

$$y[n] = \mathbf{w}^H \mathbf{x}_n, \tag{2.22}$$

where the $N \times 1$ complex vector $\mathbf{w}$ contains the beamformer weights w_k, $k = 0, 1, \ldots, N$.

The array's response to the incoming signals can be controlled by adjusting the elements of the weight vector $\mathbf{w}_n$. This process is referred to as *spatial filtering* or *beamforming*. Many methods exist to design or compute $\mathbf{w}_n$ such that it produces a desired pattern. A beampattern is a plot of the gain of the beamformer in each possible direction.

Figure 2.3 shows a plot of the beampattern of a 10-element array where all the weights are equal to 1, for angles of arrival ranging from $-90°$ to $+90°$.

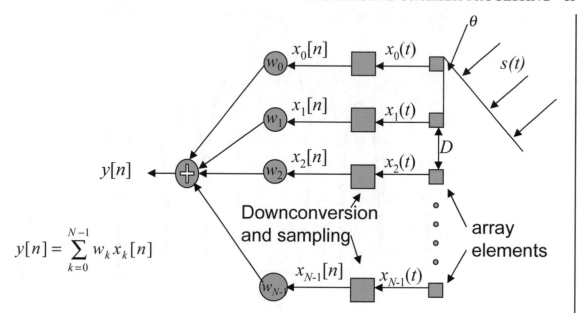

FIGURE 2.2: Narrowband beamforming system.

The magnitudes and weights of the beamformer determine the pattern and directivity of the antenna. Figure 2.2 and (2.10) can be used to write the output of the beamformer in the case that a single signal is present with angle of arrival θ.

$$y[k] = \frac{1}{N}\sum_{n=0}^{N-1} w_n x_n[k] = \frac{1}{N}\sum_{n=0}^{N-1} w_n s_0[k] e^{-j2\pi nd \sin\theta} = s_0[k]\frac{1}{N}\sum_{n=0}^{N-1} w_n e^{-j2\pi nd \sin\theta}. \quad (2.23)$$

The signal $s_0[k]$ is scaled by the following function of θ:

$$W(\theta_0) = \frac{1}{N}\sum_{n=0}^{N-1} w_n e^{-jn\omega}, \quad \text{and } \omega = 2\pi d \sin\theta, \quad (2.24)$$

where $W(\theta)$ is known as the *array factor* or *beampattern*. The beampattern can be written in vector notation as follows:

$$W(\theta_0) = \frac{1}{N}\sum_{n=0}^{N-1} w_n e^{-jn\omega} = \mathbf{w}^H \mathbf{a}(\theta), \quad (2.25)$$

where

$$\mathbf{w} = \frac{1}{N}[w_0^* \, w_1^* \, \cdots \, w_{N-1}^*]^T, \quad (2.26)$$

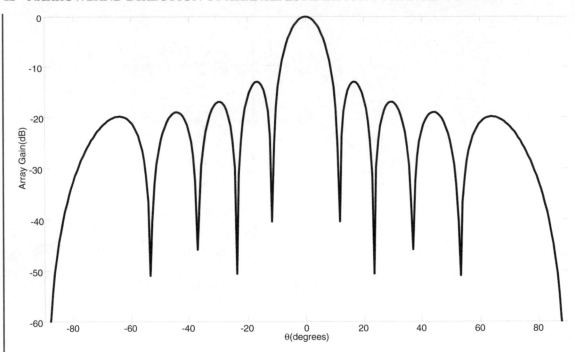

FIGURE 2.3: Polar plot of beampattern for a 10-element uniform linear array; $D = \lambda/2$.

is the vector containing the beamformer weights. The steering vector is defined as

$$\mathbf{a}(\theta) = \left[1\, e^{-j\omega}\, e^{-j2\omega}\, \ldots\, e^{-j(N-1)\omega} \right]^{\mathrm{T}}. \tag{2.27}$$

The beamformer output can also be written as a vector inner product as follows:

$$y[n] = \frac{1}{N} \sum_{k=0}^{N-1} w_k x_k[n] = \mathbf{w}^{\mathrm{H}} \mathbf{x}_n = s_0[k] \mathbf{w}^{\mathrm{H}} \mathbf{a}(\theta_0) = s_0[k] W(\theta_0). \tag{2.28}$$

The beamformer gain of the signal $s_0[k]$ is the beampattern evaluated at the angle of arrival corresponding to $s_0[k]$. The beampattern gain can be evaluated as the vector inner product of the weight vector $\mathbf{w}$ and $\mathbf{a}(\theta)$. If $\mathbf{w}$ and $\mathbf{a}(\theta)$ are orthogonal, then $\mathbf{w}^{\mathrm{H}}\mathbf{a}(\theta) = 0$ and hence the signal $s_0[k]$ is cancelled or nulled. Now, suppose two digitally modulated signals, $s_0(t)$ and $s_1(t)$, are present with angles of arrival θ_0 and θ_1, respectively. The beamformer output is given by

$$y[k] = s_0[k] W(\theta_0) + s_1[k] W(\theta_1) = s_0[k] \mathbf{w}^{\mathrm{H}} \mathbf{a}(\theta_0) + s_1[k] \mathbf{w}^{\mathrm{H}} \mathbf{a}(\theta_1). \tag{2.29}$$

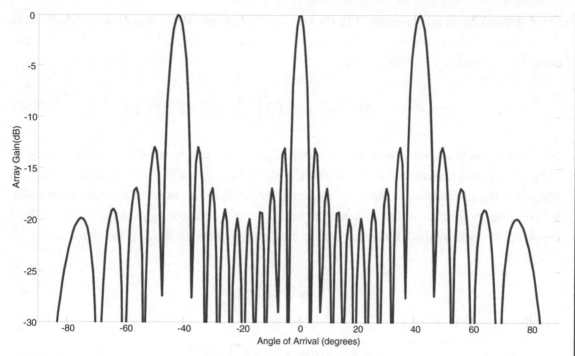

FIGURE 2.4: Demonstration of spatial aliasing with an eight-element uniform linear array with $D = 1.5\lambda$.

If $s_0(t)$ is the signal of interest and $s_1(t)$ represents interference, $\mathbf{w}$ can be designed such that $\mathbf{w}^H \mathbf{a}(\theta_0) = 1$ and $\mathbf{w}^H \mathbf{a}(\theta_1) = 0$. The output is then

$$y[k] = s_0[k]W(\theta_0) + s_1[k]W(\theta_1) = s_0[k] . \tag{2.30}$$

To avoid spatial aliasing in array processing, the element spacing must be at most $\lambda/2$ of the wavelength. If $D > 0.5\lambda$, aliasing occurs which manifests itself in terms of *grating lobes*. To illustrate spatial aliasing, we simulate a linear array with eight elements that are spaced 1.5λ apart. The beamformer weights are all equal to 1 and the beam pattern is shown in Figure 2.4. Because of aliasing, the array cannot distinguish between signals at $0°$, $-42°$, and $+42°$.

2.2.1   The Conventional Beamformer

In the case of the conventional beamformer, all the weights are given a magnitude of $1/N$ but each with a different phase, i.e.,

$$\mathbf{w} = \frac{1}{N}\,\mathbf{a}(\theta) = \frac{1}{N} \left[1 \quad e^{-j\omega} \quad e^{-j2\omega} \ldots e^{-j(N-1)\omega} \right]^{T} . \tag{2.31}$$

For a uniform linear array (Figure 1.3), the relationship between θ and ω is given in (2.24). Note that in the case where only one signal is present ($r = 1$) and neglecting noise, the beamformer output using the conventional beamformer is:

$$y_k = \mathbf{w}^H \mathbf{x}_k = s_0[k]\mathbf{w}^H \mathbf{a}(\theta) = s_0[k] \left(\frac{1}{N}\right) \mathbf{a}(\theta)^H \mathbf{a}(\theta) = s_0[k]. \tag{2.32}$$

In this case, the beamformer is said to be phase-aligned with the signal of interest and the signal of interest appears undistorted at the output. The signal-to-noise ratio (SNR) gain of the conventional beamformer can be computed by comparing the SNR at the output of a single element with the overall beamformer output. Assuming $E[|v_n[k]|^2] = \sigma^2$ and knowing $|a_n(\theta)|^2 = 1$, the SNR of the signal received by the kth element can be computed by first considering the signal model for element k.

$$x_k[n] = a_k(\theta)s_0[k] + v_k[n]. \tag{2.33}$$

The signal power is given by

$$E_s = E\left[\left|a_k(\theta)s_0[k]\right|^2\right] = E\left[\left|s_0[k]\right|^2\right], \tag{2.34}$$

and the noise power is

$$E_n = E\left[\left|v_n[k]\right|^2\right] = \sigma^2. \tag{2.35}$$

The SNR of the received signal by a single element is $(E_s/E_n)=(E_s/\sigma^2)$.

Assuming $E[\mathbf{v}_k\mathbf{v}_k^H] = \sigma^2\mathbf{I}$ and knowing $\mathbf{a}(\theta)^H\mathbf{a}(\theta) = N$, the SNR of the signal at the beamformer output can be computed by first examining the beamformer output, i.e.,

$$y_k = \mathbf{w}^H(\mathbf{a}(\theta)s_0[k] + \mathbf{v}_k) = s_0[k] + \mathbf{w}^H\mathbf{v}_k. \tag{2.36}$$

The signal power is given by,

$$E_s = E\left[\left|s_0[k]\right|^2\right] \tag{2.37}$$

and the noise power is given by,

$$E_n = E\left[\left|\mathbf{w}^H\mathbf{v}_k\right|^2\right] = \frac{1}{N^2}E\left[\mathbf{a}(\theta)^H\mathbf{v}_k\mathbf{v}_k^H\mathbf{a}(\theta)\right] = \frac{N\sigma^2}{N^2}. \tag{2.38}$$

Hence, the SNR for the overall beamformer output is $(E_s/E_n)=(NE_s/\sigma^2)$. The SNR has increased by a factor of N over the SNR at the output of a single element.

2.2.2 The Minimum Variance Distortionless Response Beamformer

The minimum variance distortionless response (MVDR) [2] beamformer is designed by minimizing the output power of the beamformer while constraining the gain to be one in the direction of interest. This problem can be stated as follows:

$$\min_{h} E[y^*y] \quad \text{subject to} \quad \mathbf{w}^H \mathbf{a}(\theta) = W(\theta) = 1. \tag{2.39}$$

The weights of the MVDR [2] are given by

$$\mathbf{w}_{\text{MVDR}} = \frac{\mathbf{R}_{xx}^{-1}\mathbf{a}(\theta)}{\mathbf{a}^H(\theta)\mathbf{R}_{xx}^{-1}\mathbf{a}(\theta)}, \tag{2.40}$$

where $\mathbf{a}(\theta)$ is the steering vector corresponding to the desired signal and $\mathbf{w}$ is the vector of complex weights. This beamformer represents a significant improvement over the conventional beamformer because, for a given DOA, it minimizes the power from unwanted directions.

CHAPTER 3

Nonadaptive Direction of Arrival Estimation

3.1 CLASSICAL METHODS FOR DIRECTION OF ARRIVAL ESTIMATION

Classical direction of arrival (DOA) methods are essentially based on beamforming. The two classical techniques for DOA are the *delay-and-sum* method and the minimum variance distortionless response (MVDR) [2] method. The basic idea behind the classical methods is to scan a beam through space and measure the power received from each direction. Directions from which the largest amount of power is received are taken to be the DOAs.

3.1.1 Delay-and-Sum Method

The delay-and-sum method computes the DOA by measuring the signal power at each possible angle of arrival and selecting as the estimate of the angle of arrival the direction of maximum power [8]. The power from a particular direction is measured by first forming a beam in that direction and setting the beamformer weights equal to the steering vector corresponding to that particular direction. The output power of the beamformer using this method is given by:

$$P(\theta) = E[\mathbf{y}^H\mathbf{y}] = E|\mathbf{w}^H\mathbf{x}_n|^2 = E|\mathbf{a}(\theta)^H\mathbf{x}_n|^2 = \mathbf{a}(\theta)^H\mathbf{R}_{xx}\mathbf{a}(\theta). \tag{3.1}$$

$P(\theta)$ will have peaks when $\mathbf{w}$ is equal to the steering vectors corresponding to the incoming signals. The disadvantage of this method is that the only way to improve the DOA resolution is to increase the number of antenna elements in the array. As was previously mentioned, the classical methods are inferior to the high-resolution subspace techniques because they do not make use of the subspace decomposition described in Chapter 2. In a linear array, the elements of the steering vectors have gains of equal magnitude, the weight vector $\mathbf{w}$ produces a sinc beampattern that has large sidelobes (see Figure 3.1). In fact, the beampattern has the same shape as the discrete time Fourier transform (DTFT) of a rectangular window. The largest sidelobe has a magnitude that is only 13 dB below that of the mainlobe. Despite the narrow mainlobe width, the large sidelobes

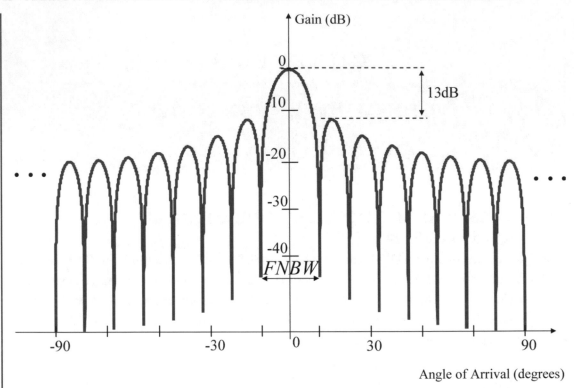

FIGURE 3.1: Beam associated with equal magnitude gains for a linear array.

allow unwanted power to enter into the computation of $P(\theta)$ for different angles of arrival and hence DOA resolution deteriorates. This method uses all the degrees of freedom to choose the weight vector with the narrowest possible beam in the direction from which the power is to be measured [15].

3.1.2   Capon's Minimum Variance Distortionless Response Method

Capon's minimum variance or MVDR was proposed in [3]. This method is similar to the delay-and-sum technique described in the previous section in that it measures the power of the received signal in all possible directions. The power from the DOA, θ, is measured by constraining the beamformer gain to be 1 in that direction and using the remaining degrees of freedom to minimize the contributions to the output power from signals coming from all other directions. The problem can be stated mathematically as a constrained minimization process. The idea is that for each possible angle, θ, the signal power in (3.2) is minimized with respect to $\mathbf{w}$ subject to the constraint that $\mathbf{w}^H \mathbf{a}(\theta) = 1$.

$$\min_{\mathbf{w}} E[|y(k)|^2] = \min_{\mathbf{w}} \mathbf{w}^H \mathbf{R} \mathbf{w}. \tag{3.2}$$

The angles for which (3.2) has peaks represent estimates of the angles of arrival of the signals. The solution to the constrained optimization problem is known as the MVDR beamformer [16] and its weights are given by:

$$\mathbf{w} = \frac{\mathbf{R}^{-1}\mathbf{a}(\theta)}{\mathbf{a}(\theta)\mathbf{R}^{-1}\mathbf{a}(\theta)}. \tag{3.3}$$

The disadvantage of this method is that an inverse matrix computation is required which may become ill-conditioned if highly correlated signals are present. This method, however, provides higher resolution than the delay-and-sum beamformer. A simulation of the MVDR and delay-and-sum methods is shown is Figure 3.2. In this simulation, a 10-element uniform linear array is used

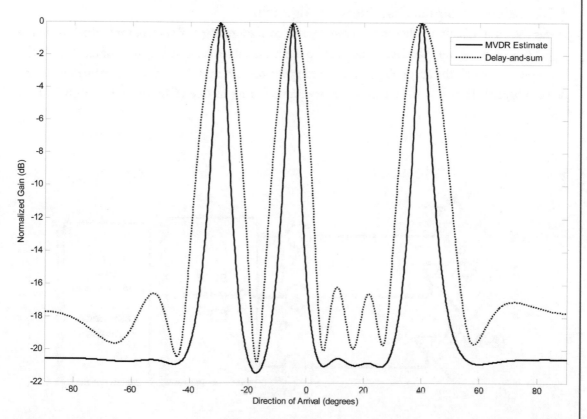

FIGURE 3.2: DOA estimation simulation for a 10-element uniform linear array. The MVDR estimator is plotted with a solid line and delay-and-sum with a dotted line. Three signals are present and the SNR is 0 dB. $D = \lambda/2$.

with a half-wavelength spacing. Three signals with equal power are present and it is clear that the MVDR method (solid line) offers superior performance. In this simulation, the data vectors were generated using (2.12) and $\mathbf{R}_{xx}$ was computed using (2.20). Three signals were present, and (3.1) and (3.2) are plotted for angles between $-90°$ and $+90°$.

3.2 SUBSPACE METHODS FOR DOA ESTIMATION

In this section, DOA estimators that make use of the signal subspace are described. These DOA estimators have high-resolution estimation capabilities. Signal subspace methods originated in spectral estimation [6] research where the autocorrelation (or autocovariance) of a signal plus noise model is estimated and then used to form a matrix whose eigenstructure gives rise to the signal and noise subspaces. The same technique can be used in array antenna DOA estimation by operating on the spatial covariance matrix as shown Figure 3.3.

3.2.1 Multiple Signal Classification Algorithm

It was shown in Chapter 2 that the steering vectors corresponding to the incoming signals lie in the signal subspace and are therefore orthogonal to the noise subspace. One way to estimate the DOAs is to search through the set of all possible steering vectors and find those that are orthogonal to the noise subspace. If $\mathbf{a}(\theta)$ is the steering vector corresponding to one of the incoming signals, then

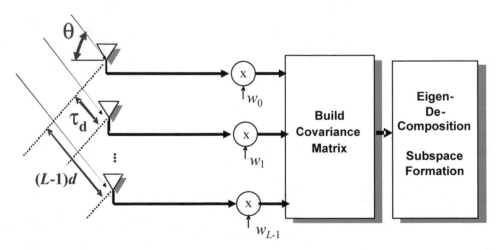

FIGURE 3.3: Eigendecomposition of antenna array signals. θ is the angle of arrival; D is the distance between two adjacent elements in meters; τ_d is the time delay of arrival between two successive elements in seconds; and there are L elements in the array.

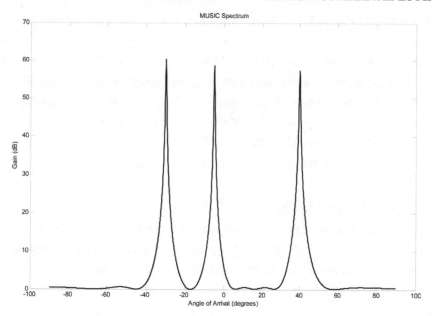

FIGURE 3.4: The MUSIC spectrum using a 10-element uniform linear array with three signals present each with an SNR of 0 dB. $d = \lambda/2$.

$\mathbf{a}(\theta)^H \mathbf{Q}_n = 0$. In practice, $\mathbf{a}(\theta)$ will not be precisely orthogonal to the noise subspace due to errors in estimating $\mathbf{Q}_n$. However the function

$$P_{\mathbf{MUSIC}}(\theta) = \frac{1}{\mathbf{a}^H(\theta)\mathbf{Q}_n \mathbf{Q}_n^H \mathbf{a}(\theta)} \tag{3.4}$$

will assume a very large value when θ is equal to the DOA of one of the signals. The above function is known as the multiple signal classification (MUSIC) "spectrum" (Figure 3.4). The MUSIC algorithm, proposed by Schmidt [11], first estimates a basis for the noise subspace, $\mathbf{Q}_n$, and then determines the r peaks in (3.4); the associated angles provide the DOA estimates.

The MUSIC algorithm has good performance and can be used with a variety of array geometries. The disadvantage of the MUSIC algorithm is that it is not able to identify DOAs of correlated signals and is computationally expensive because it involves a search over the function P_{MUSIC} for the peaks. *Spatial smoothing* can be introduced to overcome this problem. In fact, spatial smoothing is essential in a multipath propagation environment. To perform spatial smoothing, the array must be divided up into smaller, possibly overlapping subarrays and the spatial covariance matrix of each subarray is averaged to form a single, spatially smoothed covariance matrix. The MUSIC algorithm is then applied on the spatially smoothed matrix.

The following simulation illustrates the effect of spatial smoothing as used with the MUSIC algorithm. Consider the typical array antenna scenario depicted in Figure 1.3 or for the planar antenna array shown in Figure 1.4. A simulation where the receiving planar antenna array consists of 64 elements placed in an 8 × 8 grid is performed. Consider that three correlated signals arrive at the array from different directions due to multipath propagation. The direct line of sight signal arrives with a 0-dB SNR, whereas the two reflected signals arrive with −3 dB and −5 dB SNRs, respectively. Figure 3.5 shows the two-dimensional (2-D) MUSIC spectrum without spatial smoothing, which has a peak at the angle of arrival of the direct line of sight signal. The angles of arrival of the two reflected signals are not clear. In Figure 3.6, a simulation of the MUSIC algorithm with spatial smoothing is shown. In this case, the three signals are distinct.

3.2.2 Orthogonal Vector Methods

Consider a vector, $\mathbf{u}$, orthogonal to the columns of the matrix $\mathbf{A}$ (i.e., $\mathbf{u}^H\mathbf{A} = \mathbf{0}$), which implies that $\mathbf{u}$ lies in the noise subspace, $\mathbf{u} = [u_0^*, u_1^*, \ldots, u_{N-1}^*]^T$. Because the inner product of $\mathbf{u}$ with any of the r columns of $\mathbf{A}$ is zero and the structure of the columns of $\mathbf{A}$ for the uniform linear array is known, the following expansion of the inner product can be written as

$$\mathbf{u}^H \mathbf{a}(\theta_i) = u_0 + u_1 e^{-j\omega_i} + u_2 e^{-j2\omega_i} + \cdots + u_{N-1} e^{-j(N-1)\omega_i} = 0 \ . \tag{3.5}$$

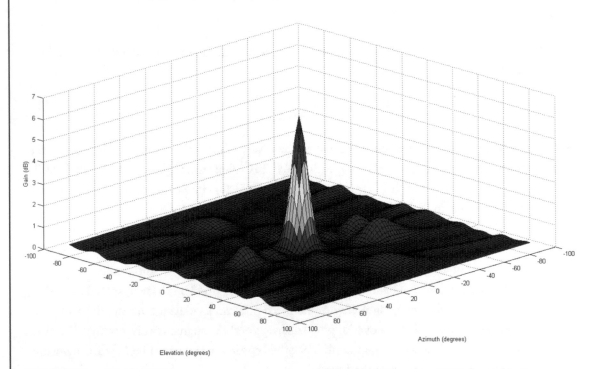

FIGURE 3.5: The MUSIC spectrum using an 8 × 8 element planar array with three signals present where no spatial smoothing is used; $D = \lambda/2$. Only one signal is detected.

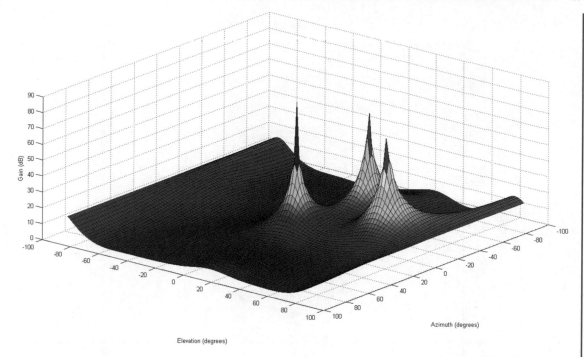

FIGURE 3.6: The MUSIC spectrum using an 8×8 element planar array with three signals present where spatial smoothing is used. $D = \lambda/2$.

Define the polynomial $u(z)$ as:

$$u(z) = u_0 + u_1 z + u_2 z^2 + \cdots + u_{N-1} z^{N-1} \tag{3.6}$$

Equations (3.5) and (3.6) indicate that the polynomial $u(z)$ evaluated at $\exp(-j\omega_i)$ is zero for $i = 0, 1, \ldots, r-1$. Therefore, the r of the roots of $u(z)$ lie on the unit circle (i.e., they all have magnitude equal to 1). Because the angles, ω_i, of these roots are functions of the DOAs (recall that $\omega_i = 2\pi d \sin(\theta_i)$), the roots of $u(z)$ can be used to compute θ_i.

To summarize the orthogonal vector methods, any vector that lies in the noise subspace is first computed. Next, a polynomial is formed whose coefficients are the elements of that vector. The r roots of the polynomial that lie on the unit circle are computed and used to determine the DOAs. This method does not work well when the SNR is low but high-performance methods based on this idea are available. Figure 3.7 shows the roots of a polynomial for a uniform linear array with 10 elements. There are three signals and nine noise components, with an overall SNR of 20 dB. The orthogonal vector methods based on this idea include Pisarenko's algorithm [29], the root MUSIC [20], and the Min-Norm [4] techniques, which are discussed next.

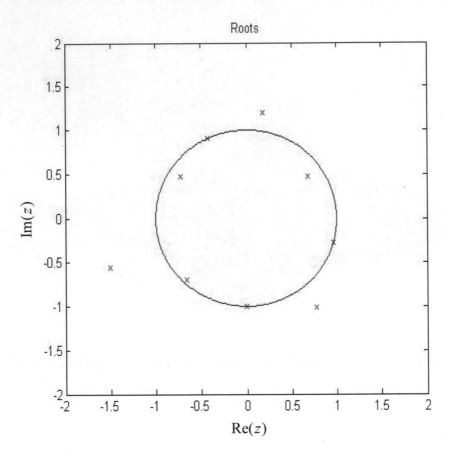

FIGURE 3.7: Roots of a polynomial whose coefficients are elements of a vector in the noise subspace with SNR = 20 dB. Three signals are present corresponding to the three roots that lie on the unit circle.

3.2.3 The Root MUSIC Algorithm

The root MUSIC algorithm was proposed by Barabell [20] and is only applicable for uniform linear arrays. It has been shown in [20] that the root MUSIC algorithm provides improved resolution relative to the ordinary MUSIC method especially at low SNRs. The steering vector for an incoming signal can be written again as defined in (2.10) and (2.13), i.e.,

$$a_n(\theta) = \exp(j2\pi nd \sin(\theta)), \quad n = 0, 1, 2,..., N-1, \tag{3.7}$$

where d is the spacing between the elements in wavelengths and θ is the angle of arrival. As was the case before, the MUSIC spectrum is defined as:

$$P_{\mathbf{MUSIC}} = \frac{1}{\mathfrak{a}^H(\theta)\mathbf{Q}_n\mathbf{Q}_n^H\,\mathbf{a}(0)} = \frac{1}{\mathfrak{a}^H(\theta)\mathbf{Ca}(0)}, \tag{3.8}$$

where $\mathbf{C}$ is

$$\mathbf{C} = \mathbf{Q}_n\mathbf{Q}_n^H. \tag{3.9}$$

By writing out the denominator as a double summation, one obtains [15]:

$$P_{\mathbf{MUSIC}}^{-1} = \sum_{k=0}^{N-1}\sum_{p=0}^{N-1} \exp(-j2\pi pd\sin\theta)\,C_{kp}\,\exp(j2\pi kd\sin\theta) \tag{3.10}$$

$$P_{\mathbf{MUSIC}}^{-1} = \sum_{p-k\,=\,\text{constant}\,=\,l} C_l\exp(-j2\pi(p-k)d\sin\theta). \tag{3.11}$$

C_l is the sum of the lth diagonal of the matrix $\mathbf{C}$. A polynomial $D(z)$ can now be defined as follows:

$$D(z) = \sum_{l=-N+1}^{N+1} C_l z^{-l}. \tag{3.12}$$

The polynomial $D(z)$ is equivalent to P_{MUSIC}^{-1} evaluated on the unit circle. Because the MUSIC spectrum will have r peaks, P_{MUSIC}^{-1} will have r valleys and hence $D(z)$ will have r zeros on the unit circle. The rest of the zeros of $D(z)$ will be away from the unit circle. It can be shown [15] that if $z_1 = be^{j\psi}$ is a root of $D(z)$, then

$$be^{j\psi} = e^{j\pi l\sin(\theta)} \Rightarrow b = 1, \tag{3.13}$$

where

$$\theta = \sin^{-1}\left(\frac{\psi_i}{l\pi}\right), \quad i = 1, 2, \ldots, d. \tag{3.14}$$

In the absence of noise, $D(z)$ will have roots that lie precisely on the unit circle, but with noise, the roots will only be close to the unit circle. The root MUSIC reduces estimation of the DOAs to finding the roots of a $(2N+1)$th-order polynomial.

3.2.4 The Minimum Norm Method

The minimum norm method was proposed by Kumaresan and Tufts [18]. This method is applied to the DOA estimation problem in a manner similar to the MUSIC algorithm. The minimum norm vector is defined as the vector lying in the noise subspace whose first element is one having minimum norm [17]. This vector is given by:

$$\mathbf{g} = \begin{bmatrix} 1 \\ \hat{\mathbf{g}} \end{bmatrix}. \tag{3.15}$$

Once the minimum norm vector has been identified, the DOAs are given by the largest peaks of the following function [17]:

$$P_{MN}(\theta) = \frac{1}{\left| \mathbf{a}^H(\theta) \begin{bmatrix} 1 \\ \hat{\mathbf{g}} \end{bmatrix} \right|}. \tag{3.16}$$

The objective now is to determine the minimum norm vector $\mathbf{g}$. Let $\mathbf{Q}_s$ be the matrix whose columns form a basis for the signal subspace. $\mathbf{Q}_s$ can be partitioned as [17]:

$$\mathbf{Q}_s = \begin{bmatrix} \alpha^* \\ \bar{\mathbf{Q}}_s \end{bmatrix}. \tag{3.17}$$

Since the vector $\mathbf{g}$ lies in the noise subspace, it will be orthogonal to the signal subspace, $\mathbf{Q}_s$, so we have the following equation [17]:

$$\mathbf{Q}_s^H \begin{bmatrix} 1 \\ \hat{\mathbf{g}} \end{bmatrix} = \mathbf{0}. \tag{3.18}$$

The above system of equations will be underdetermined, therefore we will use the minimum Frobenius norm [17] solution given by:

$$\hat{\mathbf{g}} = -\bar{\mathbf{Q}}_s \left(\bar{\mathbf{Q}}_s^H \bar{\mathbf{Q}}_s \right)^{-1} \alpha. \tag{3.19}$$

From (3.18), we can write:

$$\mathbf{I} = \mathbf{Q}_s^H \mathbf{Q}_s = \alpha \alpha^* - \bar{\mathbf{Q}}_s^H \bar{\mathbf{Q}}_s. \tag{3.20}$$

From this equation, we can write:

$$\mathbf{I} = \left(\bar{\mathbf{Q}}_s^H \bar{\mathbf{Q}}_s \right)^{-1} \alpha = (\mathbf{I} - \alpha \alpha^*)^{-1} \alpha = \alpha / (1 - \| \alpha \|^2). \tag{3.21}$$

Using (3.21), we can eliminate the calculation of the matrix inverse in (3.19). We can compute $\mathbf{g}$ based only on the orthonormal basis of the signal subspace as follows:

$$\hat{\mathbf{g}} = -\bar{\mathbf{Q}}_s \alpha / (1 - \|\alpha\|^2) . \tag{3.22}$$

Once $\mathbf{g}$ has been computed, the Min-Norm function given above is evaluated and the angles of arrival are given by the r peaks (see Figure 3.8). The Min-Norm technique is generally considered to be a high-resolution method although it is still inferior to the MUSIC and estimation of signal parameters via rotational invariance techniques (ESPRIT) algorithms.

A simulation is performed with 10 sensors in a linear array tracking three signals, each with an SNR of 0 dB. The sensors are placed half a wavelength apart. Comparative performance results using the MUSIC algorithm, the Capon algorithm, the Min-Norm algorithm, and the classical beamformer are shown in Figure 3.9. It can be seen that the MUSIC algorithm and the Capon method identify the three signals and have no other spurious components. Of the two, the MUSIC algorithm is able to better represent the locations with more prominent peaks. The Min-Norm algorithm also identifies the signals similar to the MUSIC algorithm, but produces spurious peaks at other locations. The low-resolution classical beamformer identifies the three signals, but the locations are not represented by sharp peaks, due to spectral leakage. The classical beamformer also produces several spurious peaks.

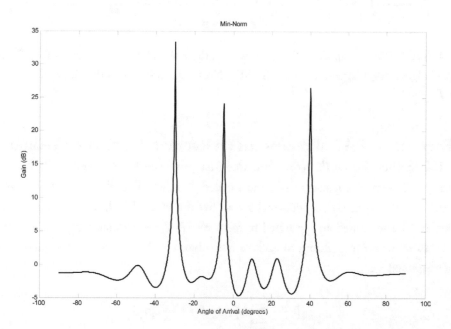

FIGURE 3.8: The Min-Norm spectrum using a 10-element uniform linear array with three signals present each with an SNR of 0 dB. $D = \lambda/2$.

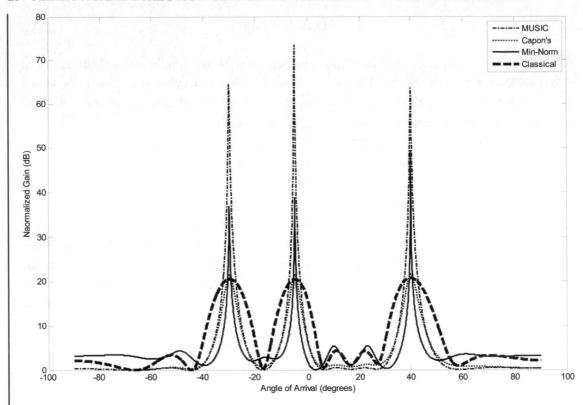

FIGURE 3.9: A 10-element uniform linear array with three signals present, each with an SNR of 0 dB. The MUSIC algorithm, Capon's method, the Min-Norm algorithm, and the classical beamformer are compared. $D = \lambda/2$.

3.2.5   Estimation of Signal Parameters via Rotational Invariance Techniques

The ESPRIT method for DOA estimation was first proposed by Roy and Kailath [9]. Assume that the array of N sensors consists of $N/2$ pairs called doublets. The displacement vector from one sensor in the doublet to its pair is identical for all the doublets. The first and second members of the doublets can be separated and grouped to form two $N/2$ element subarrays. The vectors $\mathbf{x}$ and $\mathbf{y}$ are the data vectors corresponding to each of the subarrays. The output of the subarrays $\mathbf{x}$ and $\mathbf{y}$ [9] can be expressed as:

$$x_k[n] = \sum_{i=0}^{r-1} s_i[n] a_k(\theta_i) + v_k^{(x)}[n], \tag{3.23a}$$

$$y_k[n] = \sum_{i=1}^{r-1} s_i[n] e^{j2\pi\Delta \sin\theta_k} a_k(\theta_i) + v_k^{(y)}[n], \tag{3.23b}$$

where similar notation to (2.11) has been used and Δ is the magnitude of the displacement in wavelengths from one member of each doublet to its pair. The angle of arrival estimated by the ESPRIT algorithm will be with respect to the displacement vector. The outputs of the two subarrays, $\mathbf{x}$ and $\mathbf{y}$, can be written in matrix form as follows [9]:

$$
\begin{aligned}
\mathbf{x}_n &= \mathbf{A}\mathbf{s}_n + \mathbf{v}_n^{(x)} \\
\mathbf{y}_n &= \mathbf{A}\boldsymbol{\Phi}\mathbf{s}_n + \mathbf{v}_n^{(y)}.
\end{aligned}
\tag{3.24}
$$

The matrix $\boldsymbol{\Phi}$ is a diagonal $r \times r$, whose diagonal elements are $\{\exp(j2\pi\Delta\sin\theta_0),$ $\exp(j2\pi\Delta\sin\theta_1), \ldots \exp(j2\pi\Delta\sin\theta_{r-1})\}$. Its diagonal elements are complex exponentials representing the phase delay of each of the r signals between the doublet pairs [9].

The data vectors from the two subarrays can be concatenated in the following way to form a single $2N - 2$ data vector [9] where:

$$
\mathbf{z}_n = \begin{bmatrix} \mathbf{x}_n \\ \mathbf{y}_n \end{bmatrix} = \mathbf{A}_b\mathbf{s}_n + \mathbf{v}_n
\tag{3.25a}
$$

$$
\mathbf{A}_b = \begin{bmatrix} \mathbf{A} \\ \mathbf{A}\boldsymbol{\Phi} \end{bmatrix} \quad \mathbf{v}_n = \begin{bmatrix} \mathbf{v}_n^{(x)} \\ \mathbf{v}_n^{(y)} \end{bmatrix}.
\tag{3.25b}
$$

The columns of $\mathbf{A}_b$ above will span the signal subspace of the concatenated subarrays. If $\mathbf{V}_s$ is a matrix whose columns are a basis for the signal subspace corresponding to the data vector $\mathbf{z}_n$, $\mathbf{A}_b$ and $\mathbf{V}_s$ can be related through an $r \times r$ transformation $\mathbf{T}$ given by

$$
\mathbf{V}_s = \mathbf{A}_b\mathbf{T},
\tag{3.26}
$$

and can be partitioned as follows:

$$
\mathbf{V}_s = \begin{bmatrix} \mathbf{E}_x \\ \mathbf{E}_y \end{bmatrix} = \begin{bmatrix} \mathbf{A}\mathbf{T} \\ \mathbf{A}\boldsymbol{\Phi}\mathbf{T} \end{bmatrix}.
\tag{3.27}
$$

From this partition, we see that the range or space spanned by $\mathbf{E}_x$, $\mathbf{E}_y$, and $\mathbf{A}$ is the same. Because $\mathbf{E}_x$ and $\mathbf{E}_y$ have the same range, we can define a rank r matrix $\mathbf{E}_{xy}$ [9] as follows:

$$
\mathbf{E}_{xy} = \begin{bmatrix} \mathbf{E}_x & \mathbf{E}_y \end{bmatrix}.
\tag{3.28}
$$

We now find an $r \times 2r$ rank r matrix that spans the null space of $\mathbf{E}_{xy}$. Let us call this matrix $\mathbf{F}$, and write the following equation.

$$
\mathbf{0} = \begin{bmatrix} \mathbf{E}_x & \mathbf{E}_y \end{bmatrix} \mathbf{F} = \mathbf{E}_x \mathbf{F}_x + \mathbf{E}_y \mathbf{F}_y = \mathbf{A}\mathbf{T}\mathbf{F}_x + \mathbf{A}\boldsymbol{\Phi}\mathbf{T}\mathbf{F}_y.
\tag{3.29}
$$

Let us also define $\boldsymbol{\Psi}$ as:

$$
\boldsymbol{\Psi} = -\mathbf{F}_x [\mathbf{F}_y]^{-1}.
\tag{3.30}
$$

Rearranging (3.29) gives:

$$\mathbf{E}_x \, \boldsymbol{\Psi} = \mathbf{E}_y \, .$$

(3.31)

Substituting (3.27) in (3.31) gives [9] we get

$$\mathbf{A}\mathbf{T}\boldsymbol{\Psi} = \mathbf{A}\boldsymbol{\Phi}\mathbf{T} \quad \Rightarrow \mathbf{A}\mathbf{T}\boldsymbol{\Psi}\mathbf{T}^{-1} = \mathbf{A}\boldsymbol{\Phi} \quad \Rightarrow \mathbf{T}\boldsymbol{\Psi}\mathbf{T}^{-1} = \boldsymbol{\Phi} \, .$$

(3.32)

The above equation implies that the eigenvalues of $\boldsymbol{\Psi}$ are equal to the diagonal elements of $\boldsymbol{\Phi}$. Once the eigenvalues, λ, of $\boldsymbol{\Phi}$ have been computed, the angles of arrival can be calculated using

$$\lambda_k = e^{j2\pi\Delta \sin\theta_k}$$

(3.33a)

$$\theta_k = \arcsin\left(\frac{\arg(\lambda_k)}{2\pi\Delta}\right).$$

(3.33b)

If $\mathbf{A}$ is a full-rank matrix, then the eigenvalues of the matrix $\boldsymbol{\Psi}$ are the diagonal elements of $\boldsymbol{\Phi}$ and the eigenvectors of $\boldsymbol{\Psi}$ are the columns of $\mathbf{T}$. In practice, the signal subspace is not known exactly; we only have an estimate from the sample covariance matrix $\mathbf{R}_{xx}$ or from a subspace tracking algorithm. Therefore, (3.31), $\mathbf{E}_x\boldsymbol{\Psi} = \mathbf{E}_y$, will not be exactly satisfied and we will have to resort to a least squares solution to compute $\boldsymbol{\Psi}$. The least squares process assumes that the columns in $\mathbf{E}_x$ are known exactly whereas the data in $\mathbf{E}_y$ is noisy. In this problem, this is not the case and therefore using the least squares process gives a biased estimate of $\boldsymbol{\Psi}$. If the assumption is made that $\mathbf{E}_x$ and $\mathbf{E}_y$ are equally noisy, the total least squares (TLS) criterion can be used to solve (3.31), which gives better results [9]. The algorithm is summarized in Figure 3.10.

3.2.6 Linear Prediction

Linear prediction has been used in spectral analysis and speech processing [33, 34]. Linear prediction can also be used to calculate the angle of arrival of propagating plane waves. This is done by choosing one of the sensors as the reference. We then attempt to predict the output of the reference sensor at each time instant by forming a linear combination of the outputs of the rest of the sensors. At any time n, an error can be defined as the difference between the output of the reference sensor, $x_0(n)$, and the linear combination of the signal from the rest of the sensors, $x_1(n), x_2(n), \ldots, x_{N-1}(n)$, i.e.,

$$e(n) = x_0(n) - \sum_{k=1}^{N-1} w_k x_k(n).$$

(3.34)

To find an optimal predictor, we use mean square error minimization, i.e.,

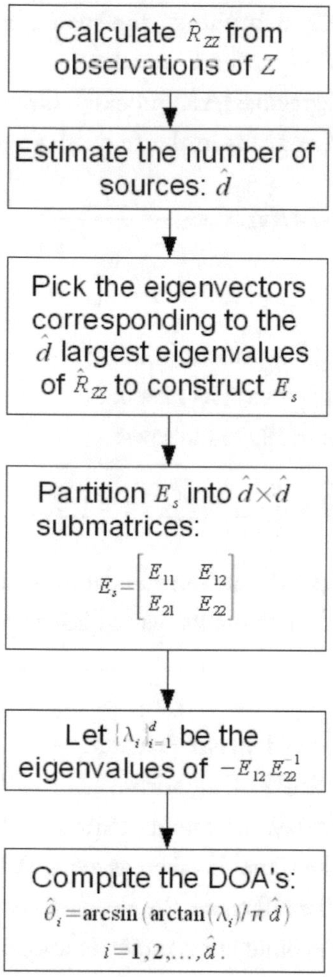

FIGURE 3.10: The ESPRIT algorithm after [9].

$$\frac{dE\left[|e(n)|^2\right]}{da_k} = E\left[-2e^*(n)x_k(n)\right], \quad k = 1, 2, \ldots, N-1 .$$

(3.35)

The above derivative is set to zero and the equation is solved for the w_k's.

$$E\left[x_l(n)x_0(n) - \sum_{k=1}^{N} w_k x_k(n)x_l(n)\right] = 0, \quad l = 1, 2, \ldots, N-1 .$$

(3.36)

The solution to this set of equations is given by

$$\mathbf{w} = \mathbf{R}_{xx}^{-1}\,\mathbf{r}_{xd} ,$$

(3.37)

where $\mathbf{R}_{xx}$ is the spatial covariance matrix and $\mathbf{r}_{xd}$ the spatial covariance vector. As in the case of spectral analysis, the linear prediction method provides coefficients for an all-pole filter (in time series spectral estimation, an autoregressive (AR) process). Once $\mathbf{w}$ has been computed, the DOAs can be determined by identifying the peaks in the frequency response of the all-pole filter whose transfer function is:

$$H(z) = \frac{1}{1 - \sum\limits_{k=1}^{N} w_k z^{-k}} . \tag{3.38}$$

It is assumed that the number of signals present is known beforehand to be r. If the r largest peaks in the above function are located at $z_i = \exp(j\phi_i)$ $i = 0, 2, ..., r - 1$, then the angles of arrival of the r signals can be related to the peaks in $H(z)$ as follows:

$$\theta_i = \frac{\pi}{2}\sin^{-1}(\phi_i), \quad i = 1, 2, ..., r. \tag{3.39}$$

The linear prediction method works for a uniform linear array and could also be extended to work with a planar array. More information about the use of linear prediction for DOA estimation can be found in [21].

3.2.7   The Unitary ESPRIT for Linear Arrays

The unique feature of the unitary ESPRIT algorithm [22] is that it can operate with strictly real computations. In a uniform linear array, the center element of the array can be taken as the reference element where the phase of the signal is taken as zero. When the number of elements is odd and the center element is taken as the reference, the steering vector for the uniform linear array will be conjugate centrosymmetric (i.e., conjugate symmetric about the center element). This steering vector [22] is given by:

$$\mathbf{a}_N(\theta) = \left[e^{-j\frac{(N-1)}{2}\omega} \;\cdots\; e^{-j\omega} \; 1 \; e^{j\omega} \;\cdots\; e^{j\frac{(N-1)}{2}\omega} \right]^{\mathrm{T}}, \tag{3.40}$$

where $\omega = 2\pi d \sin(\theta)$. When N is even, the reference point of the array is taken as the center of the array even though no element is positioned there, i.e.,

$$\mathbf{a}_N(\theta) = \left[e^{-j\frac{N}{2}\omega} \ldots e^{-j\omega/2} \; e^{j\omega/2} \; e^{j3\omega/2} \ldots e^{j\frac{N}{2}\omega} \right]^{\mathrm{T}}. \tag{3.41}$$

Define the matrix $\prod_N$ as the $N \times N$ matrix with ones on the antidiagonal and zeros elsewhere. Using this matrix, the following relationship can be established:

$$\Pi_N \, \mathbf{a}_N(\theta) = \mathbf{a}_N^*(\theta), \tag{3.42}$$

where $*$ denotes conjugation of the matrix elements. The conjugate centrosymmetric steering vector $\mathbf{a}_N(\theta)$ can be transformed to a real vector through a unitary transformation whose rows are centrosymmetric. One possible transformation when N is even and $N = 2k$ is:

$$\mathbf{Q}_N = \frac{1}{\sqrt{2}} \begin{bmatrix} \mathbf{I}_k & j\mathbf{I}_k \\ \Pi_k & -j\Pi_k \end{bmatrix}. \tag{3.43}$$

If N is odd, $N = 2k + 1$ and

$$\mathbf{Q}_N = \frac{1}{\sqrt{2}} \begin{bmatrix} \mathbf{I}_k & \mathbf{0} & j\mathbf{I}_k \\ \mathbf{0}^T & \sqrt{2} & \mathbf{0}^T \\ \Pi_k & \mathbf{0} & -j\Pi_k \end{bmatrix}. \tag{3.44}$$

The centroconjugate symmetric steering vector $\boldsymbol{a}_N(\theta)$ can be transformed to a real vector $\mathbf{d}_N(\theta)$ as follows:

$$\mathbf{d}_N(\theta) = \mathbf{Q}_N^H \mathbf{a}_N(\theta) = [\cos\left(u\tfrac{N-1}{2}\right) \dots \cos(u) \tfrac{1}{\sqrt{2}}\cos(0) -\sin\left(u\tfrac{N-1}{2}\right) \dots$$
$$\dots - \sin(u)]^T. \tag{3.45}$$

Next, the covariance matrix of the transformed received array data is given by [22]:

$$\hat{\mathbf{R}}_{yy} = E\left[\mathbf{y}\mathbf{y}^H\right] \quad \text{where} \quad \mathbf{y} = \left(\mathbf{Q}_N^H\right)\mathbf{x}. \tag{3.46}$$

Let us now examine the effect of transforming the data vectors by $\mathbf{Q}_N$ just as the steering vectors were transformed in (3.45). We assume that the data vectors obey the model described in (2.13), $\mathbf{x} = \mathbf{As} + \mathbf{v}$, where the columns of $\mathbf{A}$ are the steering vectors of the incoming signals. Transforming $\mathbf{x}$ gives:

$$\mathbf{y} = \mathbf{Q}_N^H\mathbf{x} = \mathbf{Q}_N^H\mathbf{As} + \mathbf{Q}_N^H\mathbf{v} = \mathbf{Ds} + \mathbf{Q}_N^H\mathbf{v}$$
$$= \mathbf{D}\,\mathrm{Re}\{\mathbf{s}\} + j\mathbf{D}\,\mathrm{Im}\{\mathbf{s}\} + \mathrm{Re}\left\{\mathbf{Q}_N^H\mathbf{v}\right\} + j\mathrm{Im}\left\{\mathbf{Q}_N^H\mathbf{v}\right\}. \tag{3.47}$$

The columns of matrix $\mathbf{D}$ are the real valued transformed steering vectors. From the equation above, one can see that in the absence of noise, $\mathbf{y}$ will simply be a linear combination of the columns of the matrix $\mathbf{D}$. Therefore, the columns of $\mathbf{D}$ span the transformed signal subspace. This signal subspace can be estimated by either taking the real part of the transformed received array data vectors and finding a basis for the signal subspace of that set or operating in the same manner on the imaginary part. Both sets of data share a common signal subspace. Alternatively, the real and imaginary vectors can be combined into one large set of vectors and the signal subspace can be computed for the combined set. This allows all of the processing to be done with real valued computations [22].

If the first $N-1$ elements of $\mathbf{a}_N(\theta)$ are multiplied by $e^{j\omega}$, the resulting $(N-1) \times 1$ vector will be equal to the last $N-1$ components of $\mathbf{a}_N(\theta)$. This can be expressed mathematically as:

$$e^{j\omega}\mathbf{J}_1\mathbf{a}_N(\theta) = \mathbf{J}_2\mathbf{a}_N(\theta), \tag{3.48}$$

where $\mathbf{J}_1$ is an $(N-1) \times N$ matrix constructed by taking the first $N-1$ rows of the $N \times N$ identity matrix and $\mathbf{J}_2$ is the $(N-1) \times N$ matrix constructed by taking the last $N-1$ rows of the $N \times N$ identity matrix. The relation in the previous equation is known as the *invariance relation* [22]. Because $\mathbf{Q}_N$ is unitary, the following can be written:

$$e^{j\omega}\mathbf{J}_1\mathbf{Q}_N^H\mathbf{Q}_N\mathbf{a}_N(\theta) = \mathbf{J}_2\mathbf{Q}_N^H\mathbf{Q}_N\mathbf{a}_N(\theta). \tag{3.49}$$

Now, using the definition [22] of $\mathbf{d}_N(\theta)$ in (3.45), we obtain:

$$e^{j\omega}\mathbf{J}_1\mathbf{Q}_N^H\mathbf{d}_N(\theta) = \mathbf{J}_2\mathbf{Q}_N^H\mathbf{d}_N(\theta). \tag{3.50}$$

Using the following identities,

$$\Pi_{N-1}\mathbf{J}_2\Pi_N = \mathbf{J}_1, \qquad \Pi_N\Pi_N = \mathbf{I}, \tag{3.51}$$

and multiplying $\mathbf{J}_2$ by Π_{N-1} on the left flips $\mathbf{J}_2$ up and down, and multiplying by Π_N on the right flips it left to right, resulting in the matrix $\mathbf{J}_1$.

$$\begin{aligned}\mathbf{Q}_{N-1}^H\mathbf{J}_2\mathbf{Q}_N &= \mathbf{Q}_{N-1}^H\Pi_{N-1}\Pi_{N-1}\mathbf{J}_2\Pi_N\Pi_N\mathbf{Q}_N \\ &= \mathbf{Q}_{N-1}^T\mathbf{J}_1\mathbf{Q}_N^* = \left(\mathbf{Q}_{N-1}^H\mathbf{J}_1\mathbf{Q}_N\right)^*.\end{aligned} \tag{3.52}$$

The above equation uses the fact that $\Pi_N\mathbf{Q}_N = \mathbf{Q}_N^*$ and $\mathbf{Q}_{N-1}^H\ \Pi_{N-1} = \mathbf{Q}_{N-1}^T$. Now, let $\mathbf{K}_1$ and $\mathbf{K}_2$ be the real and imaginary parts of $\mathbf{Q}_{N-1}^H\mathbf{J}_2\mathbf{Q}_N$, respectively. If we multiply the above equation by $\mathbf{Q}_{N-1}^H$, we obtain:

$$e^{j\omega}\mathbf{Q}_{N-1}^H\mathbf{J}_1\mathbf{Q}_N\,\mathbf{d}_N(\theta) = \mathbf{Q}_{N-1}^H\mathbf{J}_2\mathbf{Q}_N\,\mathbf{d}_N(\theta). \tag{3.53}$$

Using the definitions for $\mathbf{K}_1$ and $\mathbf{K}_2$, the above equation becomes

$$e^{j\omega}(\mathbf{K}_1 - j\mathbf{K}_2)\mathbf{d}_N(\theta) = (\mathbf{K}_1 + j\mathbf{K}_2)\mathbf{d}_N(\theta) \tag{3.54}$$

$$e^{j\omega/2}(\mathbf{K}_1 - j\mathbf{K}_2)\mathbf{d}_N(\theta) = e^{-j\omega/2}(\mathbf{K}_1 + j\mathbf{K}_2)\mathbf{d}_N(\theta). \tag{3.55}$$

Now rearrange by grouping the $\mathbf{K}_1$ and $\mathbf{K}_2$ terms

$$\left(e^{j\omega/2} - e^{-j\omega/2}\right) \mathbf{K}_1 \mathbf{d}_N(\theta) = \left(e^{j\omega/2} + e^{-j\omega/2}\right) j\mathbf{K}_2 \mathbf{d}_N(\theta). \tag{3.56}$$

Using trigonometric identities we get:

$$\tan(\omega/2)\mathbf{K}_1 \mathbf{d}_N(\theta) = \mathbf{K}_2 \mathbf{d}_N(\theta). \tag{3.57}$$

Now suppose that the DOAs are $\{\theta_1, \theta_2, \ldots, \theta_d\}$. Now, (3.57) can be extended to include all of $\mathbf{d}_N(\theta)$ as follows:

$$\mathbf{K}_1 \mathbf{D}\Omega = \mathbf{K}_2 \mathbf{D}, \tag{3.58}$$

where [22]

$$\Omega = \text{diag}\{\tan(\pi d \sin\theta_1) \ldots \tan(\pi d \sin\theta_d)\}. \tag{3.59}$$

The columns of $\mathbf{D}$ are the transformed steering vectors corresponding to the r incoming signals. As shown at the very beginning of this discussion, the signal subspace estimated from the real and imaginary vectors of the transformed data space $\mathbf{y}$ will span the same space spanned by the columns of the matrix $\mathbf{D}$. If the basis for the signal subspace is contained in the columns of the matrix $\mathbf{E}_s$, then the matrices $\mathbf{D}$ and $\mathbf{E}_s$ can be related by a matrix $\mathbf{T}$. $\mathbf{E}_s = \mathbf{DT}$, where $\mathbf{T}$ is $r \times r$. Substituting $\mathbf{D} = \mathbf{E}_s\mathbf{T}^{-1}$ into the equation $\mathbf{K}_1\mathbf{D}\Omega = \mathbf{K}_2\Omega$, the result becomes:

$$\mathbf{K}_1\mathbf{E}_s\mathbf{T}^{-1}\Omega = \mathbf{K}_2\mathbf{E}_s\mathbf{T}^{-1} \tag{3.60}$$

or

$$\mathbf{K}_1\mathbf{E}_s\mathbf{T}^{-1}\Omega\mathbf{T} = \mathbf{K}_2\mathbf{E}_s. \tag{3.61}$$

Let

$$\Psi = \mathbf{T}^{-1}\Omega\mathbf{T} \tag{3.62}$$

then the above equation becomes [22]

$$\mathbf{K}_1\mathbf{E}_s\Psi = \mathbf{K}_2\mathbf{E}_s. \tag{3.63}$$

This says that the eigenvalues of Ψ are $\tan(d\pi\theta_i)$, $i = 1, 2, \ldots, r$, and Ψ can be computed as the least squares solution to $\mathbf{K}_1\mathbf{E}_s\Psi = \mathbf{K}_2\mathbf{E}_s$. This can be done because $\mathbf{K}_1$ and $\mathbf{K}_2$ are known and $\mathbf{E}_s$ can be estimated from the data. Once $\mathbf{E}_s$ is estimated, the DOAs can be computed [22] as the eigenvalues, λ_i, of Ψ, i.e.,

$$\lambda_i = \tan(\pi d \sin \theta_i) \qquad \theta_i = \tan^{-1}\left[(\lambda_i)/(\pi d)\right]. \tag{3.64}$$

Note that the estimation of the matrix $\mathbf{E}_s$ and the solutions (3.63) and (3.64) involve only real computations.

A summary of the algorithm is shown in Figure 3.11.

3.2.8   QR ESPRIT

The Total Least Squares (TLS) ESPRIT involves computing a singular value decomposition, a matrix inverse, a matrix product, and an eigenvalue decomposition of an $r \times r$ matrix. This is a heavy computational burden, especially if DOAs are to be tracked across time. An alternative to the above procedure is QR reduction to a standard eigenvalue problem [6]. One can start with the generalized eigenvalue problem that is associated with the ESPRIT algorithm, $\mathbf{E}_x\mathbf{T}\Phi = \mathbf{E}_y\mathbf{T}$. In this case, $\mathbf{E}_x$ is a matrix

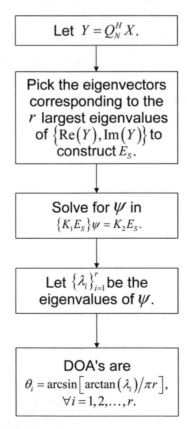

FIGURE 3.11: The unitary ESPRIT [22].

whose columns represent the signal subspace of the first subarray, and the matrix $\mathbf{E}_y$ contains the signal subspace for the second of the two subarrays. Next, premultiply the above equation for the generalized eigenvalue problem by the matrix $\mathbf{T}^H\mathbf{E}_x^H$:

$$\mathbf{T}^H\,\mathbf{E}_x^H\,\mathbf{E}_x\mathbf{T}\boldsymbol{\Phi} = \mathbf{T}^H\,\mathbf{E}_x^H\,\mathbf{E}_x\mathbf{T}. \tag{3.65}$$

According to Strobach [6], the matrix $\mathbf{T}$ can be chosen such that

$$\mathbf{T}^H\,\mathbf{E}_x^H\,\mathbf{E}_x\mathbf{T} = \mathbf{I}. \tag{3.66}$$

If $\mathbf{T}$ is chosen to satisfy (3.66), then

$$\boldsymbol{\Phi} = \mathbf{T}^H\,\mathbf{E}_x^H\,\mathbf{E}_y\mathbf{T}. \tag{3.67}$$

Now suppose that $\mathbf{E}_x = \mathbf{Q}_x\mathbf{R}_x$ and $\mathbf{Q} = \mathbf{R}_x\mathbf{T}$. Then, (3.65) can be written as $\mathbf{Q}_x\mathbf{R}_x\mathbf{T}\boldsymbol{\Phi} = \mathbf{E}_x\mathbf{R}_x^{-1}\mathbf{R}_x\mathbf{T}$, i.e.,

$$\mathbf{Q}_x\mathbf{Q}\boldsymbol{\Phi} = \mathbf{E}_x^H\mathbf{R}_x^{-1}\mathbf{Q} \quad \Rightarrow \quad \mathbf{Q}\boldsymbol{\Phi} = \mathbf{Q}_x^H\mathbf{E}_x^H\mathbf{R}_x^{-1}\mathbf{Q}, \tag{3.68}$$

or

$$\mathbf{Q}\boldsymbol{\Phi}\mathbf{Q}^{-1} = \mathbf{Q}_x^H\,\mathbf{E}_x^H\,\mathbf{R}_x^{-1}. \tag{3.69}$$

Equation (3.69) is an eigenvalue problem in standard form. To summarize, the DOAs can be computed via a QR reduction, which corresponds to the solution of the ESPRIT problem but not the TLS ESPRIT. Figure 3.12 is a flowchart of what could be called the QR ESPRIT algorithm [6].

3.2.9 Beamspace DOA Estimation

Beamspace algorithms are efficient in terms of computational complexity. They use an $N \times P$ beamspace matrix, $\mathbf{T}$, whose columns represent the beamformer weights. With this approach the data vectors are transformed to a lower dimensional space by the matrix $\mathbf{T}$. This transformation is written as, $\mathbf{z}_n = \mathbf{T}^H\mathbf{x}_n$. The P elements of the vector $\mathbf{z}_n$ can be thought of as outputs of P beamformers. If $\mathbf{z}_n = \mathbf{T}^H\mathbf{x}_n^H$, then the DOA algorithm operates on the transformed data space contained in the columns of $\mathbf{z}_n$. If information on the incoming signal direction is available, the columns of $\mathbf{T}$ can be designed such that beams in the columns of $\mathbf{T}$ point in the general direction of the signals whose DOAs are to be estimated. In block DOA estimation, the eigendecomposition of an $N \times N$ matrix requires $O(N^3)$ operations. If two beamspace processors are designed, $\mathbf{T}_1$ and $\mathbf{T}_2$, such that $\mathbf{T}_1$ covers from $0°$ to $90°$ and $\mathbf{T}_2$ covers from $0°$ to $-90°$, then the two beamspace processors can estimate the

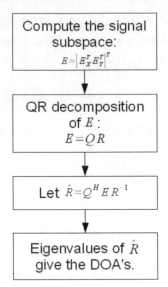

FIGURE 3.12: The QR ESPRIT algorithm [6].

DOAs in their respective sectors. Therefore, the number of computations can be reduced considerably as each beamspace processor will have complexity $O((N/2)^3)$. Several subspace-based DOA algorithms have been developed for beamspace processors such as the beamspace MUSIC [23], the beamspace root MUSIC [24], the beamspace ESPRIT [25], and the DFT beamspace ESPRIT [22]. In this section, the ESPRIT versions of the beamspace DOA algorithms are described.

3.2.10 The DFT Beamspace ESPRIT

Recall that for a uniform linear array, the conjugate symmetric steering vector has the form

$$\mathbf{a}(\theta) = \left[e^{-j(\frac{N-1}{2})2\pi d \sin\theta} \quad \ldots \quad e^{-j2\pi d \sin\theta} \quad 1 \quad e^{j2\pi d \sin\theta} \quad \ldots \quad e^{j(\frac{N-1}{2})2\pi d \sin\theta} \right]. \qquad (3.70)$$

Now consider the inner product [22] of the above steering vector with the mth row of the centrosymmetrized DFT matrix, which is given by:

$$\mathbf{w}_m^H = e^{-j(\frac{N-1}{2})m\frac{2\pi}{N}} \left[1 \quad e^{-j\frac{2\pi}{N}m} \quad e^{-j2\frac{2\pi}{N}m} \quad \ldots \quad e^{-j(N-1)m\frac{2\pi}{N}} \right]. \qquad (3.71)$$

Note that this is a scaled version of the mth row of the DFT matrix. The inner product of $\mathbf{w}_m$ and $\mathbf{a}(\theta)$ is

$$\mathbf{w}_m^H \mathbf{a}(\theta) = \frac{\sin\left(\frac{N}{2}\left(2\pi d \sin\theta - m\frac{2\pi}{N}\right)\right)}{\sin\left(\frac{1}{2}\left(2\pi d \sin\theta - m\frac{2\pi}{N}\right)\right)} = \mathbf{b}_m(\theta). \qquad (3.72)$$

Now let $\mathbf{b}_N(\theta)$ be an $N \times 1$ vector containing the N samples of the centrosymmetrized DFT of the steering vector $\mathbf{a}(\theta)$, i.e.,

$$\mathbf{b}_N(\theta) = \begin{bmatrix} b_1(\theta) \; b_2(\theta) \; \ldots \; b_N(\theta) \end{bmatrix}^{\mathrm{T}}. \tag{3.73}$$

Notice that the numerator of $b_m(\theta)$ and $b_{m+1}(\theta)$ are the negative of one another since the arguments of the sine waves are π radians apart. This observation leads to the following equation [22]

$$b_m(\mu)\, \sin\left[\frac{1}{2}\left(\mu - m\frac{2\pi}{N}\right)\right] + b_{m+1}(\mu)\, \sin\left[\frac{1}{2}\left(\mu - (m+1)\frac{2\pi}{N}\right)\right] = 0, \tag{3.74}$$

where $\mu = 2\pi d \sin\theta$. Now, using the trigonometric identity $\sin(\theta - \phi) = \sin(\theta)\cos(\phi) - \sin(\phi)\cos(\theta)$, the above equation can be written as

$$\tan\left(\frac{\mu}{2}\right)\left(b_m(\mu)\,\cos\left(\frac{m\pi}{N}\right) + b_{m+1}(\mu)\,\cos\left(\frac{(m+1)\pi}{N}\right)\right) = b_m(\mu)\sin\left(\frac{m\pi}{N}\right)$$

$$+ b_{m+1}(\mu)\sin\left(\frac{\pi(m+1)}{N}\right). \tag{3.75}$$

The previous two equations can be written in matrix notation but to do this, first it is necessary to relate $b_0(\theta)$ to $b_{N-1}(\theta)$. Let us first write $b_N(\theta)$ as follows:

$$b_N(\theta) = \frac{\sin\left(\frac{N}{2}\left(\mu - N\frac{2\pi}{N}\right)\right)}{\sin\left(\frac{1}{2}\left(\mu - N\frac{2\pi}{N}\right)\right)} = \frac{\sin\left(\frac{N}{2}\mu - N\pi\right)}{\sin\left(\frac{1}{2}\mu - \pi\right)} = \frac{(-1)^N \sin\left(\frac{N}{2}\mu\right)}{-\sin\left(\frac{1}{2}\mu\right)}$$

$$= (-1)^{N-1} b_0(\mu). \tag{3.76}$$

Now use the above equation along with the equation relating $b_m(\theta)$ to $b_{m+1}(\theta)$ with $m = N - 1$ to establish the following relationship between $b_0(\theta)$ and $b_{N-1}(\theta)$:

$$\tan\left(\frac{\mu}{2}\right)\left(b_{N-1}(\mu)\cos\left(\frac{(N-1)\pi}{N}\right) + (-1)^{N-1} b_0(\mu)\,\cos(\pi)\right)$$

$$= b_{N-1}(\mu)\sin\left(\frac{(N-1)\pi}{N}\right) + (-1)^{N-1} b_0(\mu)\,\sin\left(\frac{\pi(N+1)}{N}\right). \tag{3.77}$$

As per [22], these equations can be used to write a matrix equation relating the first $N-1$ elements of $\mathbf{b}_N(\theta)$ to the last N elements of $\mathbf{b}_N(\theta)$., i.e.,

$$\tan\left(\frac{\mu}{2}\right)\begin{bmatrix} 1 & \cos\left(\dfrac{\pi}{N}\right) & 0 & 0 & \cdots & 0 & 0 & 0 \\[2mm] 0 & \cos\left(\dfrac{\pi}{N}\right) & \cos\left(\dfrac{2\pi}{N}\right) & 0 & \cdots & 0 & 0 & 0 \\[2mm] 0 & 0 & 0 & 0 & \cdots & 0 & 0 & 0 \\[1mm] \vdots & \vdots & \vdots & \vdots & \ddots & \vdots & \vdots & \vdots \\[1mm] 0 & 0 & 0 & 0 & \cdots & 0 & \cos\left(\dfrac{(N-2)\pi}{N}\right) & \cos\left(\dfrac{(N-1)\pi}{N}\right) \end{bmatrix}\mathbf{b}_N(\mu)$$

$$=\begin{bmatrix} 0 & \sin\left(\dfrac{\pi}{N}\right) & 0 & 0 & \cdots & 0 & 0 & 0 \\[2mm] 0 & \sin\left(\dfrac{\pi}{N}\right) & \sin\left(\dfrac{2\pi}{N}\right) & 0 & \cdots & 0 & 0 & 0 \\[2mm] 0 & 0 & 0 & 0 & \cdots & 0 & 0 & 0 \\[1mm] \vdots & \vdots & \vdots & \vdots & \ddots & \vdots & \vdots & \vdots \\[1mm] 0 & 0 & 0 & 0 & \cdots & 0 & \sin\left(\dfrac{(N-2)\pi}{N}\right) & \sin\left(\dfrac{(N-1)\pi}{N}\right) \end{bmatrix}\mathbf{b}_N(\mu),$$

$$\tan\left(\frac{\mu}{2}\right)\Gamma_1\mathbf{b}_N(\mu) = \Gamma_2\mathbf{b}_N(\mu). \tag{3.78}$$

Now if r signals are present, then the transformed steering vectors are given by: $\mathbf{b}_N(\theta_0)$, $\mathbf{b}_N(\theta_1)$, ..., $\mathbf{b}_N(\theta_{r-1})$; if a matrix $\mathbf{B}$ is formed using columns that are the transformed steering vectors, then this equation can be written as

$$\Gamma_1\mathbf{B}\Omega = \Gamma_2\mathbf{B}, \quad \Omega = \operatorname{diag}\left\{\tan\frac{\mu_0}{2}, \tan\frac{\mu_1}{2}, \ldots, \tan\frac{\mu_{r-1}}{2}\right\}. \tag{3.79}$$

The transformed steering vectors and the signal subspace of the transformed data vectors should span approximately the same subspace. Therefore, for some $r \times r$ matrix $\mathbf{T}$, $\mathbf{E}_s = \mathbf{BT}$ and hence $\mathbf{B} = \mathbf{E}_s\mathbf{T}^{-1}$. Substituting this equation in (3.79), we obtain

$$\Gamma_1\mathbf{E}_s\mathbf{T}^{-1}\Omega = \Gamma_2\mathbf{E}_s\mathbf{T}^{-1}, \quad \Gamma_1\mathbf{E}_s\Psi = \Gamma_2\mathbf{E}_s, \quad \text{with} \quad \Psi = \mathbf{T}^{-1}\Omega\mathbf{T}. \tag{3.80}$$

In Figure 3.13, we summarize the steps of the DFT beamspace ESPRIT [22].

3.2.11 The Multiple Invariance ESPRIT
It can be shown that the TLS ESPRIT algorithm is based on the following minimization problem [32]:

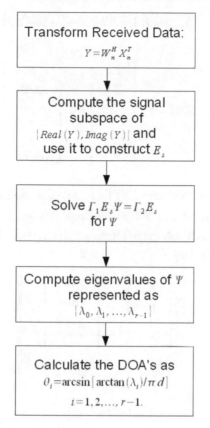

FIGURE 3.13: The DFT beamspace ESPRIT [22].

$$\min_{\{\mathbf{B},\mathbf{\Psi}\}} \left\| \begin{bmatrix} \hat{\mathbf{E}}_0 \\ \hat{\mathbf{E}}_1 \end{bmatrix} - \begin{bmatrix} \mathbf{B} \\ \mathbf{B}\mathbf{\Psi} \end{bmatrix} \right\|, \tag{3.81}$$

with $\mathbf{B} = \mathbf{AT}$ and $\mathbf{\Psi} = \mathbf{T}^{-1}\mathbf{\Phi}\mathbf{T}$. We know that the actual signal subspace $\mathbf{E}_s$ spans the same subspace as the columns of the steering matrix $\mathbf{A}$; therefore, the two matrices could be related through a rotation matrix $\mathbf{T}$, i.e., $\mathbf{E}_s = \mathbf{AT}$. In the ESPRIT algorithm, a single invariance is exploited. A selection matrix is $\mathbf{J} = [\mathbf{J}_0^T \mathbf{J}_1^T]^T$, where $\mathbf{J}_0$ is the $(N - 1) \times N$ matrix created by taking the first $N - 1$ rows of the $N \times N$ identity matrix and similarly, $\mathbf{J}_1$ is created by taking the last $N - 1$ rows of the same identity matrix. In the case of a uniform linear array, the following equation holds:

$$\mathbf{JE}_s = \mathbf{JAT} = \begin{bmatrix} \mathbf{E}_0 \\ \mathbf{E}_1 \end{bmatrix} = \begin{bmatrix} \bar{\mathbf{A}} \\ \bar{\mathbf{A}}\mathbf{\Phi} \end{bmatrix} \mathbf{T}. \tag{3.82}$$

Now suppose that the array has more than one identical subarray where the subarrays are displaced from the reference subarray by a vector that points in the same direction but has a different length.

Let these vectors be $\Delta_0, \Delta_1, \ldots, \Delta_{p-1}$ and let $\mathbf{J}$ be a selection matrix for the subarrays with $\mathbf{J} = [\mathbf{J}_0^T \mathbf{J}_1^T \ldots \mathbf{J}_{p-1}^T]^T$.

$$\mathbf{JE}_s = \mathbf{JAT} \Rightarrow \begin{bmatrix} \mathbf{E}_0 \\ \mathbf{E}_1 \\ \cdot \\ \cdot \\ \cdot \\ \mathbf{E}_{p-1} \end{bmatrix} = \begin{bmatrix} \mathbf{A} \\ \mathbf{A\Phi} \\ \mathbf{A\Phi}^{\delta_2} \\ \cdot \\ \cdot \\ \mathbf{A\Phi}^{\delta_{p-1}} \end{bmatrix} \mathbf{T}, \tag{3.83}$$

where $\delta_i = |\Delta_i|/|\Delta_1|$. The multiple invariance ESPRIT (MI-ESPRIT) process is essentially based on a subspace fitting [30] formulation and can be posed as follows. First, a signal subspace estimate denoted $\hat{\mathbf{E}}$ is computed for each subarray. Then, we determine the matrices $\mathbf{A}$, $\mathbf{\Phi}$, and $\mathbf{T}$ that minimize the function V given below

$$V = \left\| \begin{bmatrix} \hat{\mathbf{E}}_0 \\ \hat{\mathbf{E}}_1 \\ \cdot \\ \cdot \\ \cdot \\ \hat{\mathbf{E}}_{p-1} \end{bmatrix} - \begin{bmatrix} \mathbf{A} \\ \mathbf{A\Phi} \\ \mathbf{A\Phi}^{\delta_2} \\ \cdot \\ \cdot \\ \mathbf{A\Phi}^{\delta_{p-1}} \end{bmatrix} \mathbf{T} \right\|_F^2. \tag{3.84}$$

This minimization is nonlinear and in [32] it is solved by using Newton's method. It is shown that the MI-ESPRIT algorithm outperforms ESPRIT, MUSIC, and root MUSIC, where performance is measured in terms of the root mean square error of the estimates.

3.2.12 Unitary ESPRIT for Planar Arrays

The unitary ESPRIT algorithm can be extended to two dimensions for use with a uniform rectangular array [31]. If an $M \times N$ rectangular array is used, the steering information will be contained in a matrix instead of a vector. This matrix is $\mathbf{A}(\theta_x, \theta_y)$, where θ_x is the angle of arrival with respect to the x axis and θ_y is the angle with respect to the y axis. The matrix $\mathbf{A}(\theta_x, \theta_y)$ can be transformed into a vector by stacking its columns thus forming a stacked steering vector $\mathbf{a}(\theta_x, \theta_y)$. The (m, n) elements of the matrix $\mathbf{A}(\theta_x, \theta_y)$ can be written as

$$A(\theta, \phi)_{m,n} = \exp(-j2\pi f_c \tau_{m,n}), \tag{3.85}$$

where

$$\tau_{n,m} = \frac{n d_x \sin(\theta_x) + m d_y \sin(\theta_y)}{c}, \tag{3.86}$$

and c is the speed of the wave. Note that d_x and d_y are the spacing between columns and rows of the array in wavelengths, respectively. The matrix $\mathbf{A}(\theta_x, \theta_y)$ can be written [31] as the outer product of two vectors of the form (3.40) as follows:

$$\mathbf{A}(\theta_x, \theta_y) = \mathbf{a}_M(\theta_x)\mathbf{a}_N^{\mathrm{T}}(\theta_y). \tag{3.87}$$

The matrix $\mathbf{A}(\theta_x, \theta_y)$ has complex entries; they can be made real valued by premultiplying by $\mathbf{Q}_M$ and postmultiplying by $\mathbf{Q}_N^*$. The real array manifold matrix then becomes

$$\mathbf{D}(\theta_x, \theta_y) = \mathbf{Q}_M^{\mathrm{H}}\mathbf{a}_M(\theta_x)\mathbf{a}_N^{\mathrm{T}}(\theta_y)\mathbf{Q}_N^* = \mathbf{d}_M(\theta_x)\mathbf{d}_N^{\mathrm{T}}(\theta_y). \tag{3.88}$$

For the one-dimensional case, the invariance relation is given in (3.57). If this invariance relation is multiplied on both sides by $\mathbf{d}_M^{\mathrm{T}}(\theta_y)$, the following result is obtained [31]

$$\tan(u/2)\mathbf{K}_1\mathbf{D}\left(\theta_x, \theta_y\right) = \mathbf{K}_2\mathbf{D}\left(\theta_x, \theta_y\right). \tag{3.89}$$

$\mathbf{D}(\theta_x, \theta_y)$ can be vectorized to create a vector $\mathbf{d}(\theta_x, \theta_y)$ by stacking columns of $\mathbf{D}(\theta_x, \theta_y)$. The above equation then becomes:

$$\tan(u/2)\mathbf{K}_{u1}\mathbf{d}\left(\theta_x, \theta_y\right) = \mathbf{K}_{u2}\mathbf{d}\left(\theta_x, \theta_y\right), \tag{3.90}$$

with

$$\mathbf{K}_{u1} = \mathbf{I}_M \otimes \mathbf{K}_1, \quad \mathbf{K}_{u2} = \mathbf{I}_M \otimes \mathbf{K}_2, \tag{3.91}$$

where Hadamard products are used. Similarly, $\mathbf{d}_M^{\mathrm{T}}(\theta_y)$ satisfies

$$\tan(v/2)\mathbf{K}_3\mathbf{d}(\theta_y) = \mathbf{K}_4\mathbf{d}_N(\theta_y). \tag{3.92}$$

We then multiply the above equation on the right by $\mathbf{d}_N(\theta_x)$ and we obtain:

$$\tan(v/2)\mathbf{K}_3\mathbf{D}^{\mathrm{T}}(\theta_x, \theta_y) = \mathbf{K}_4\mathbf{D}^{\mathrm{T}}(\theta_x, \theta_y). \tag{3.93}$$

If we transpose both sides

$$\tan(v/2)\mathbf{D}(\theta_x, \theta_y)\mathbf{K}_3^{\mathrm{T}} = \mathbf{D}(\theta_x, \theta_y)\mathbf{K}_4^{\mathrm{T}}. \tag{3.94}$$

The above equation can again be vectorized, i.e.,

$$\tan(v/2)\mathbf{K}_{v1}\mathbf{d}(\theta_x, \theta_y) = \mathbf{K}_{v2}\mathbf{d}(\theta_x, \theta_y), \tag{3.95}$$

and

$$\mathbf{K}_{v1} = \mathbf{K}_3 \otimes \mathbf{I}_N, \quad \mathbf{K}_{v2} = \mathbf{K}_4 \otimes \mathbf{I}_N. \tag{3.96}$$

Suppose that the incoming signals have angles of arrival: $(\theta_{x1}, \theta_{y1}), (\theta_{x2}, \theta_{y2}), \ldots, (\theta_{xd}, \theta_{yd})$. A matrix $\mathbf{D}$ can be formed from the vectors $\mathbf{d}(\theta_{x1}, \theta_{y1}), \mathbf{d}(\theta_{x2}, \theta_{y2}), \ldots, \mathbf{d}(\theta_{xd}, \theta_{yd})$. The above equation can be written using matrix $\mathbf{D}$ as

$$\mathbf{K}_{v1} \mathbf{D} \Omega_v = \mathbf{K}_{v2} \mathbf{D}, \tag{3.97}$$

where

$$\Omega_v = \mathrm{diag}\left\{ \tan\left(\pi d \sin \theta_{y1}\right), \ldots, \tan\left(\pi d \sin \theta_{yd}\right) \right\}. \tag{3.98}$$

By the same argument,

$$\mathbf{K}_{u1} \mathbf{D} \Omega_u = \mathbf{K}_{u2} \mathbf{D}, \tag{3.99}$$

and

$$\Omega_u = \mathrm{diag}\left\{ \tan(\pi d \sin \theta_{x1}), \ldots, \tan(\pi d \sin \theta_{xd}) \right\}. \tag{3.100}$$

As was done for the uniform linear array, the data vector, $\mathbf{x}$, obtained from the array may be transformed by $\mathbf{Q}_N$ and $\mathbf{Q}_M$ as follows:

$$\mathbf{y} = \mathbf{Q}_N^H \mathbf{X} \mathbf{Q}_M^*, \tag{3.101}$$

or if $\mathbf{y}$ and $\mathbf{x}$ are to be vectorized:

$$\mathbf{y} = \left(\mathbf{Q}_M^H\right) \otimes \left(\mathbf{Q}_M^H\right) \mathbf{x}. \tag{3.102}$$

The data space is now spanned by $\{\mathrm{Re}(\mathbf{y}), \mathrm{Im}(\mathbf{y})\}$ and if $\mathbf{E}_s$ is a basis for the signal subspace of the transformed data, then the columns of $\mathbf{D}$ and $\mathbf{E}_s$ span the same space and can be related through an $r \times r$ matrix $\mathbf{T}$, i.e.,

$$\mathbf{E}_s = \mathbf{D} \mathbf{T}, \quad \mathbf{D} = \mathbf{E}_s \mathbf{T}^{-1}. \tag{3.103}$$

Substituting for $\mathbf{D}$ in the invariance equations above, we get

$$\mathbf{K}_{v1} \mathbf{E}_s \mathbf{T}^{-1} \Omega_v = \mathbf{K}_{v2} \mathbf{E}_s \mathbf{T}^{-1}. \tag{3.104}$$

Let $\mathbf{\Psi}_v = \mathbf{T}^{-1}\mathbf{\Omega}_v\mathbf{T}$. The DOAs can thus be computed by solving the above least squares problem and then finding the eigenvalues of the solution matrix, which contain the information [31] about the DOA.

3.2.13 Maximum Likelihood Methods

The maximum likelihood (ML) estimator performs better than the methods discussed previously but at the cost of increased computational complexity. This method can identify DOAs even for certain types of correlated signals. Assume that there are r signals arriving at angles, which are contained in the vector $\mathbf{\theta}$,

$$\mathbf{\theta} = \begin{bmatrix} \theta_0 \, \theta_1 \, \ldots \, \theta_{r-1} \end{bmatrix}. \tag{3.105}$$

The matrix $\mathbf{X}$ is the data matrix whose ith column is $\mathbf{x}_i$. In this case [15], the joint probability density function of $\mathbf{X}$ is:

$$f(\mathbf{X}) = \prod_{i=0}^{K-1} \frac{1}{\pi \det\left(\sigma^2\mathbf{I}\right)} \exp\left(-\frac{1}{\sigma^2}|\mathbf{x}_i - \mathbf{A}(\theta_i)\mathbf{s}_i|^2\right). \tag{3.106}$$

Neglecting the constant terms, the columns of $\mathbf{A}$, or steering vectors, are functions of the elements of $\mathbf{\theta}$. The log likelihood function then becomes:

$$L = -Kd\log\sigma^2 - \frac{1}{\sigma^2}\sum_{l=0}^{K-1}|\mathbf{x}_i(l) - \mathbf{A}(\theta)\mathbf{s}_i|^2. \tag{3.107}$$

Therefore, L must now be maximized with respect to the unknown parameters $\mathbf{s}$ and θ. This is equivalent to the following minimization problem:

$$\min_{\theta_k,\mathbf{s}} \left\{\sum_{i=0}^{K-1}|\mathbf{x}_i - \mathbf{A}(\theta_i)\mathbf{s}_i|^2\right\}. \tag{3.108}$$

If $\mathbf{\theta}$ is fixed and the function in (3.108) is minimized with respect to $\mathbf{s}$, the least squares solution [15] can be written as:

$$\mathbf{s}_i = \left(\mathbf{A}^H(\theta_i)\,\mathbf{A}(\theta_i)\right)^{-1}\mathbf{A}^H(\theta)\,\mathbf{x}_i. \tag{3.109}$$

Substituting the above least squares solution in the function in (3.108), we obtain:

$$\min_{\theta_k} \left\{\sum_{i=0}^{K-1}\left|\mathbf{x}_i - \mathbf{A}(\theta_i)\left(\mathbf{A}^H(\theta_i)\mathbf{A}(\theta_i)\right)^{-1}\mathbf{A}^H(\theta_i)\mathbf{x}_i\right|^2\right\} = \min_{\theta_k}\left\{\sum_{i=1}^{L}\left|\mathbf{x}_i - \mathbf{P}_{\mathbf{A}(\theta_i)}(\theta_i)\mathbf{x}_i\right|^2\right\}. \tag{3.110}$$

The matrix $\mathbf{P}$ in equation (3.110) is a projector onto the space spanned by the columns of $\mathbf{A}$. This is equivalent to maximizing the following log likelihood function:

$$L(\boldsymbol{\theta}) = \sum_{i=0}^{K-1} |\mathbf{P}_{\mathbf{A}(\theta_i)} \mathbf{x}_i|^2. \tag{3.111}$$

This function can be maximized by finding a set of steering vectors whose span closely approximates the span of the data vectors; note that the data vectors are the rows of $\mathbf{X}$. This closeness will be measured by the magnitude of the projection of the rows of $\mathbf{X}$ onto the span of $\mathbf{A}$. The choice of $\mathbf{A}$ that provides the largest magnitude is considered to be the closest [15].

3.2.13.1 The Alternating Projection Algorithm for ML DOA Estimation

Ziskind and Wax [19] proposed an algorithm for maximizing the likelihood function in (3.107). The method is known as the *Alternating Projection Algorithm (APA)*. The APA is shown in Figure 3.14 and summarized below:

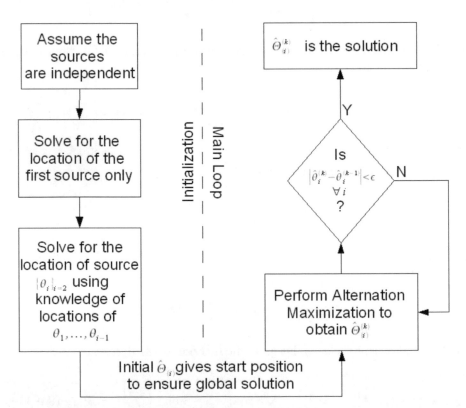

FIGURE 3.14: The Alternating Projection Algorithm (APA) [19].

Step 1. Initialization: It is important to have a good initial guess of the matrix $\mathbf{P}$ so that the algorithm avoids convergence to a local minimum. This can be done by first finding a 1-D projection, $\mathbf{P}$, that maximizes L. This projector will be constructed from a single vector from the array manifold. The angle corresponding to that vector will be the initial estimate of θ_0. This vector is denoted $\mathbf{a}^0(\theta_0)$. Now, we find the 2-D projector using $\mathbf{a}^0(\theta_0)$ and a vector from the manifold that maximizes L. This vector will be called $\mathbf{a}^0(\theta_1)$. This procedure is followed until $\mathbf{P}$ has been expanded into an r-dimensional projector. The initial estimates of the angles of arrival correspond to the steering vectors that are used to form the projection matrix $\mathbf{P}$, i.e.,

$$\left[\mathbf{a}^0(\theta_0)\ \mathbf{a}^0(\theta_1)\ \ldots\ \mathbf{a}^0(\theta_{r-1})\right]. \tag{3.112}$$

Note that the superscript refers to the iteration number.

Step 2. Next, the steering vectors are held $\mathbf{a}^0(\theta_1)$, $\mathbf{a}^0(\theta_2)$, ..., $\mathbf{a}^0(\theta_{r-1})$ fixed and we search for a new $\mathbf{a}^0(\theta_0)$ that maximizes L. This new estimate of $\mathbf{a}^0(\theta_0)$ replaces the old one and is denoted $\mathbf{a}^1(\theta_0)$. We then proceed to hold $\mathbf{a}^0(\theta_0)$, $\mathbf{a}^0(\theta_2)$, ..., $\mathbf{a}^0(\theta_{r-1})$ fixed and search for a new $\mathbf{a}^0(\theta_1)$ that maximizes L. This new estimate of $\mathbf{a}^0(\theta_1)$ will be denoted $\mathbf{a}^1(\theta_1)$. We continue in this manner until a new estimate is obtained for each $\mathbf{a}^1(\theta_i)$. This constitutes one iteration.

Step 3. Repeat Step 2 until the variation in the vectors $\mathbf{a}^k(\theta_i)$ is below a certain tolerance factor.

· · · ·

CHAPTER 4

Adaptive Direction of Arrival Estimation

The algorithms described in the previous chapters assumed that the data is stationary, which will be the case if the sources are not moving in space. When the sources are not stationary, then algorithms that continuously reestimate the direction of arrival (DOA) must be developed. For sources moving in space, the columns of the matrix **A** in (2.13) become time-varying and hence the span of the signal subspace will be changing with time. To track the signal subspace, a subspace tracking algorithm must be used. One way to develop an adaptive DOA algorithm is to concatenate a subspace tracker with a DOA algorithm. At each iteration, the subspace tracker passes an estimate of the signal or noise subspace to the DOA algorithm and it estimates the DOAs.

Adaptive versions of the estimation of signal parameters via rotational invariance techniques (ESPRIT algorithm) that can efficiently update the DOA estimates with every iteration have been developed. In [6], it was shown that equation (3.31) can be solved adaptively. Another algorithm [13] is available that adaptively and efficiently computes the eigenvalues of (3.32). One of the benchmark algorithms for subspace tracking is Karasalo's [28] subspace tracker.

Most subspace tracking algorithms are based on the $N \times N$ covariance matrix of the data and most algorithms use an exponentially weighted estimate of the covariance matrix in place of the estimator in (2.20), which is only useful when the signal sources are not moving in space. The equation for the exponentially weighted estimate at time n is

$$\mathbf{R}_{xx}(n) = \alpha \mathbf{R}_{xx}(n-1) + (1-\alpha)\mathbf{x}_n\mathbf{x}_n^{\mathrm{H}}, \tag{4.1}$$

where $0 < \alpha < 1$ is the forgetting factor. With a small forgetting factor, α, less emphasis is placed on past data and more emphasis is placed on the current data vector. A small value of α gives good tracking ability but poor steady-state accuracy, whereas a larger value of α gives slow tracking ability but provides good steady-state performance.

Karasalo's [28] subspace tracking algorithm uses the concept of deflation to reduce the dimension of the matrices involved from $N \times N$ to $(r+1) \times (r+2)$. Suppose that the eigendecomposition of the spatial correlation matrix $\mathbf{R}_{xx}(n)$ has the following structure:

$$\mathbf{R}_{xx}(n) = \mathbf{Q}(n)\mathbf{D}(n)\mathbf{Q}^{H}(n) = \begin{bmatrix} \mathbf{Q}_s(n) & \mathbf{Q}_n(n) \end{bmatrix} \begin{bmatrix} \mathbf{D}_s(n) & 0 \\ 0 & \sigma^2(n)\mathbf{I}_n \end{bmatrix} \begin{bmatrix} \mathbf{Q}_s^{H}(n) \\ \mathbf{Q}_n^{H}(n) \end{bmatrix}, \qquad (4.2)$$

where $\sigma_{r+1}(n) = \sigma_{r+2}(n) = \cdots = \sigma_N(n) = \sigma$, $\mathbf{D}_s(n) = \text{diag}\{\sigma_1, \sigma_2, \ldots, \sigma_r\}$, and the columns of the matrix $\mathbf{Q}(n) = [\mathbf{Q}_s(n)\mathbf{Q}_n(n)]$ are the eigenvectors of $\mathbf{R}_{xx}(n)$. Because $\mathbf{R}_{xx}(n)$ is positive semidefinite, the eigenvectors will form an orthogonal set and the eigenvalues will be real. The columns of $\mathbf{Q}_s(n)$ form an orthonormal basis for the signal subspace and the columns of $\mathbf{Q}_n(n)$, an orthonormal basis for the noise subspace. $\mathbf{D}_s(n)$ is a diagonal matrix that contains the signal eigenvalues of the matrix $\mathbf{R}_{xx}(n)$ and the noise eigenvalues are $\sigma^2(n)$. Therefore, all eigenvalues in the noise subspace are the same, which means that we have a *spherical subspace*. This means that any vector lying in that subspace, regardless of its orientation, will be an eigenvector with eigenvalue $\sigma^2(n)$ and therefore the matrix $\mathbf{Q}_n(n)$ can be rotated without affecting the above eigendecomposition of $\mathbf{R}_{xx}(n)$ as long as the columns of $\mathbf{Q}_n(n)$ remain orthogonal to the signal subspace. Given this freedom, $\mathbf{Q}_n(n)$ is chosen to be $[\mathbf{w}_1(n)\mathbf{C}(n)]$, where $\mathbf{w}_1(n)$ is the normalized component of $\mathbf{x}_n$ lying in the noise subspace and $\mathbf{C}(n)$ is a matrix whose columns form a basis for the subspace that is orthogonal to $\mathbf{w}_1(n)$ and to the signal subspace. The data vector, $\mathbf{x}_n$, can be decomposed into these two components as follows:

$$\mathbf{x}_n = \mathbf{Q}_s(n-1)\mathbf{z}_n + c_1(n)\mathbf{w}_1(n) \quad \text{where} \quad \mathbf{w}_1^H(n)\mathbf{w}_1(n) = 1 \quad \text{and} \quad \mathbf{Q}_s(n-1)\mathbf{w}_1(n) = 0, \qquad (4.3)$$

$$\mathbf{w}_1(n) = \frac{\mathbf{x}_n - \mathbf{Q}_s(n-1)\mathbf{Q}_s^H(n-1)\mathbf{x}_n}{\sqrt{\|\mathbf{x}_n - \mathbf{Q}_s(n-1)\mathbf{Q}_s^H(n-1)\|}}, \quad c_1(n) = \sqrt{\|\mathbf{x}_n - \mathbf{Q}_s(n-1)\mathbf{Q}_s^H(n-1)\|}, \qquad (4.4)$$

$$\mathbf{z}_n = \mathbf{Q}_s^H(n-1)\mathbf{x}_n. \qquad (4.5)$$

The first term in the decomposition of $\mathbf{x}_n$ in (4.3) lies in the signal subspace and the second term in the noise subspace. Suppose the spatial covariance matrix $\mathbf{R}_{xx}(n)$ is updated using an exponentially weighted estimate as given in (4.1), using (4.2)–(4.5) in (4.1), $\mathbf{R}_{xx}(n)$ can be written as:

$$\mathbf{R}_{xx}(n) = (1 - \alpha)\begin{bmatrix} \mathbf{Q}_s(n-1)\mathbf{z}_n + c_1(n)\mathbf{w}_1(n) \end{bmatrix}\begin{bmatrix} \mathbf{Q}_s(n-1)\mathbf{z}_n + c_1(n)\mathbf{w}_1(n) \end{bmatrix}^H$$

$$+ \alpha \begin{bmatrix} \mathbf{Q}_s(n-1) & \mathbf{Q}_n(n-1) \end{bmatrix} \begin{bmatrix} \mathbf{D}_s(n-1) & 0 \\ 0 & \sigma^2(n-1)\mathbf{I}_n \end{bmatrix} \begin{bmatrix} \mathbf{Q}_s^H(n-1) \\ \mathbf{Q}_n^H(n-1) \end{bmatrix}. \qquad (4.6)$$

The term on the right is the eigendecomposition of the spatial covariance matrix at time $(n-1)$. With some rearranging of the terms in the above equation and using $\mathbf{Q}_n(n) = [\mathbf{w}_1(n)\mathbf{C}(n)]$, (4.6) can be written as:

$$\mathbf{R}_{xx}(n) = \begin{bmatrix} \mathbf{Q}_s(n-1) & \mathbf{w}_1(n) \end{bmatrix} \begin{bmatrix} \alpha^{1/2}\mathbf{D}_s^{1/2}(n-1) & 0 & (1-\alpha)^{1/2}\mathbf{z}_n \\ 0 & \alpha^{1/2}\sigma(n-1) & (1-\alpha)^{1/2}\mathbf{c}_1(n) \end{bmatrix}$$

$$\times \begin{bmatrix} \alpha^{1/2}\mathbf{D}_s^{1/2}(n-1) & 0 \\ 0 & \alpha^{1/2}\sigma(n-1) \\ (1-\alpha)^{1/2}\mathbf{z}_n^H & (1-\alpha)^{1/2}\mathbf{c}_1(n) \end{bmatrix} \begin{bmatrix} \mathbf{Q}_s^H(n-1) \\ \mathbf{w}_1^H(n) \end{bmatrix} + \alpha\sigma^2(n-1)\mathbf{C}(n)\mathbf{C}^H(n) . \tag{4.7}$$

The columns of the matrix term on the left lie in the subspace spanned by the columns of $[\mathbf{Q}_s(n-1)\mathbf{w}_1(n)]$. The term on the right lies in the space spanned by the columns of $\mathbf{C}(n)$, which lies in the noise subspace. Therefore, the columns of the two matrix terms in (4.7) span two orthogonal subspaces. The columns of $\mathbf{C}(n)$ are eigenvectors of $\mathbf{R}_{xx}(n)$ that lie in the noise subspace. To compute the rest of the eigenvectors (the r signal eigenvectors and the remaining noise eigenvectors come from the first term), it is necessary to write the first term in (4.7) in terms of its eigendecomposition $\mathbf{U}(n)\mathbf{D}(n)\mathbf{U}^H(n)$. This can be done by first computing the singular value decomposition (SVD) of the $(r+1) \times (r+2)$ matrix

$$\mathbf{L}(n) = \begin{bmatrix} \alpha^{1/2}\mathbf{D}_s^{1/2}(n-1) & 0 & (1-\alpha)^{1/2}\mathbf{z}_n \\ 0 & \alpha^{1/2}\sigma(n-1) & (1-\alpha)^{1/2}\mathbf{c}_1(n) \end{bmatrix}. \tag{4.8}$$

Suppose its SVD is $\mathbf{V}(n)\mathbf{S}(n)\mathbf{Y}^H(n)$, then the first term in (4.7) can be written as:

$$\mathbf{R}_{xx}(n) = \begin{bmatrix} \mathbf{Q}_s(n-1) & \mathbf{w}_1(n) \end{bmatrix} \mathbf{V}(n)\mathbf{S}(n)\mathbf{Y}^H(n)\mathbf{Y}(n)\mathbf{S}^H(n)\mathbf{V}^H(n) \begin{bmatrix} \mathbf{Q}_s^H(n-1) \\ \mathbf{w}_1^H(n) \end{bmatrix}$$

$$+ \alpha\sigma^2(n-1)\mathbf{C}(n)\mathbf{C}^H(n)$$

$$= \begin{bmatrix} \mathbf{Q}_s(n-1) & \mathbf{w}_1(n) \end{bmatrix}\mathbf{V}(n)\mathbf{S}(n)\mathbf{S}^H(n)\mathbf{V}^H(n) \begin{bmatrix} \mathbf{Q}_s^H(n-1) \\ \mathbf{w}_1^H(n) \end{bmatrix}$$

$$+ \alpha\sigma^2(n-1)\mathbf{C}(n)\mathbf{C}^H(n) \tag{4.9}$$

$$= \mathbf{K}(n)\mathbf{S}(n)\mathbf{S}^H(n)\mathbf{K}^H(n) + \sigma^2(n-1)\mathbf{C}(n)\mathbf{C}^H(n)$$

$$= \begin{bmatrix} \mathbf{K}(n) & \mathbf{C}(n) \end{bmatrix} \begin{bmatrix} \mathbf{S}(n)\mathbf{S}^H(n) & 0 \\ 0 & \alpha\sigma^2(n-1)\mathbf{I}_{N-r-1} \end{bmatrix} \begin{bmatrix} \mathbf{K}^H(n) \\ \mathbf{C}^H(n) \end{bmatrix}.$$

The matrix $\mathbf{R}_{xx}(n)$ is now in its eigendecomposition form and it has been assumed that $\mathbf{K}(n) = [\mathbf{Q}_s(n-1)\mathbf{w}_1(n)]\mathbf{V}(n)$. Assume also that the SVD has been computed in such a way that the diagonal elements of the matrix, in the middle of the above product, are in descending order. Then the signal eigenvectors, or those corresponding to the largest eigenvalues of $\mathbf{R}_{xx}(n)$, are the first r columns of $\mathbf{K}(n)$ and the noise eigenvectors of $\mathbf{R}_{xx}(n)$ are the columns of $\mathbf{C}(n)$ and $(r+1)^{\text{st}}$ column of $\mathbf{K}(n)$. If the $(r+1) \times (r+1)$ matrix $\mathbf{V}(n)$ is partitioned as follows:

$$\mathbf{V}(n) = \begin{bmatrix} \boldsymbol{\theta}(n)^* \\ \mathbf{f}^{\mathrm{H}}(n)^* \end{bmatrix}, \tag{4.10}$$

where the matrix $\boldsymbol{\theta}(n)$ has dimensions $r \times r$ and the vector $\mathbf{f}^{\mathrm{H}}(n)$ has dimensions $1 \times r$, then the updated signal subspace can be written as

$$\mathbf{Q}_s(n) = \mathbf{Q}_s(n-1)\boldsymbol{\theta}(n) + \mathbf{w}_1(n)\mathbf{f}^{\mathrm{H}}(n). \tag{4.11}$$

TABLE 4.1: Summary of Karasalo's Subspace Tracking Algorithm [119, 121]

Initialization
for $n = 1, 2, ...$

$\mathbf{z}_n = \mathbf{Q}_s^{\mathrm{H}}(n-1)\mathbf{x}_n$

$\mathbf{w}_1(n) = \mathbf{x}(n) - \mathbf{Q}_s(n-1)\mathbf{z}_1(n)$

$\mathbf{c}_1(n) = \mathbf{w}_1^{\mathrm{H}}(n)\mathbf{w}_1(n)$

$\mathbf{w}_1(n) = \mathbf{w}_1(n)/\sqrt{\mathbf{c}_1(n)}$

$\mathbf{L}(n) = \begin{bmatrix} \alpha^{1/2}\mathbf{D}_s^{1/2}(n-1) & 0 & (1-\alpha)^{1/2}\mathbf{z}_n \\ 0 & \alpha^{1/2}\sigma(n-1) & (1-\alpha)^{1/2}\mathbf{c}_1(n) \end{bmatrix}$

$\mathbf{L}(n) = \mathbf{V}(n)\mathbf{S}(n)\mathbf{Y}^{\mathrm{H}}(n)$. Perform SVD of $\mathbf{L}(n)$.

$\mathbf{V}(n) = \begin{bmatrix} \boldsymbol{\theta}(n)^* \\ \mathbf{f}^{\mathrm{H}}(n)^* \end{bmatrix}$

$\mathbf{Q}_s(n) = \mathbf{Q}_s(n-1)\boldsymbol{\theta}(n) + \mathbf{w}_1(n)\mathbf{f}^{\mathrm{H}}(n)$

$\sigma^2(n) = \dfrac{1}{N-r}\sigma_{r+1}^2(n) + \alpha\dfrac{N-r-1}{N-r}\sigma^2(n-1)$

end(n)

Next, the estimate of $\sigma^2(n)$ is updated from the new eigendecomposition of $\mathbf{R}_{xx}(n)$ given in (4.9), where the diagonal elements of $\mathbf{S}(n)$ are $\sigma_1(n)$, $\sigma_2(n)$, ..., $\sigma_{r+1}(n)$ and have been placed in descending order. The $N-r$ noise eigenvalues are $\sigma^2_{r+1}(n)$, $\alpha\sigma^2(n-1)$, $\alpha\sigma^2(n-1)$, ..., $\alpha\sigma^2(n-1)$, where $\alpha\sigma^2(n-1)$ is listed $N-r-1$ times. Taking an average of these values gives

$$\sigma^2(n) = \frac{1}{N-r}\sigma^2_{r+1}(n) + \alpha\frac{N-r-1}{N-r}\sigma^2(n-1). \qquad (4.12)$$

Karasalo's method is an SVD updating algorithm and is often used as a reference method for comparing other subspace tracking algorithms. Although it is a good reference method, Karasalo's algorithm is not often used in certain practical applications because the computation of the $(r+1) \times (r+2)$ SVD at each iteration is itself an iterative process. Other similar algorithms exist that replace the $(r+1) \times (r+2)$ SVD with a more efficient adaptive method [27, 28].

4.1   ADAPTIVE SIMULATION EXAMPLE

In this simulation, there are three signal sources moving in space with time. Their DOAs are changing by 0.01° per iteration. The subspace tracking algorithm in Table 4.1 is used to track the signal subspace. The DOAs are recomputed after each iteration using the estimate of the signal subspace, $\mathbf{Q}$s, and the ESPRIT algorithm given in (3.31)–(3.33a). The dotted line in Figure 4.1 represents the true DOA and the solid line represents the estimate of the DOA by the adaptive algorithm.

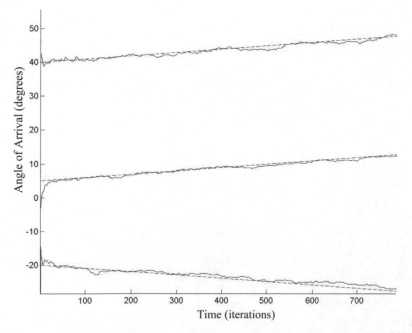

FIGURE 4.1: Adaptive DOA simulation using Karasalo's subspace tracker and the ESPRIT algorithm.

Appendix

This appendix describes a MATLAB m-file that implements four of the DOA algorithms described in this book for a uniform linear array.

SIGNAL GENERATOR

The signal generator implements the data model in equation (2.13). The steering vector matrix $\mathbf{A}$ from (2.10)–(2.12) is computed. The number of signals are specified by the number of elements in the in the vector **doas**. The vector $\mathbf{P}$ is the same length as **doas** and contains the corresponding power of the signals. Other parameters that can be set include the number elements N in the array, the distance d between elements in wavelengths, the number of data snapshots K to generate, and the variance of the uncorrelated noise present at each element.

The spatial correlation matrix, $\mathbf{R}_{xx}$, is computed by using an implementation of (2.20) that uses matrix multiplication of the data matrix $\mathbf{X}$. The eigendecomposition of $\mathbf{X}$ is performed using the MATLAB *eig* function. The eigenvectors are then sorted based on their eigenvalues. The eigenvectors corresponding to the r largest eigenvalues are used as a basis for the signal subspace $\mathbf{Q}_s$. The eigenvectors corresponding to the smallest $N - r$ eigenvalues are used as a basis for the noise subspace.

THE MUSIC ALGORITHM

The MATLAB code for the MUSIC algorithm is an implementation of (3.4). Equation (3.4) is sampled by creating an array of steering vectors corresponding to the angles in the vector **angles**. The estimate of the noise subspace computed by the signal generator is used in this computation.

THE ESPRIT ALGORITHM

The first line of the MATLAB implementation of the ESPRIT algorithm is of (3.31). $\mathbf{E}_x$ and $\mathbf{E}_y$ can be obtained by taking the first and last $N - 1$ rows, respectively, of the signal subspace matrix $\mathbf{Q}_s$. This is a more efficient way that explicitly computes the signal subspace for each subarray. Next, (3.32) and (3.33a) are implemented to compute the DOAs of the incoming signals.

MVDR METHOD AND THE CLASSICAL BEAMFORMER

The MVDR beamformer has been implemented by using (2.40) directly along with the array of steering vectors that was previously computed for the MUSIC algorithm. Finally, the delay and sum or classical beamforming method described in (3.1) is implemented for comparison.

CODE TO SIMULATE THE MUSIC, THE ESPRIT, THE MVDR, THE MIN-NORM, AND THE CLASSICAL DOA ALGORITHMS

```
% Simulation of MUSIC, ESPRIT, MVDR, Min-Norm and Classical DOA
% algorithms for a uniform linear array.

doas=[-30 -5 40]*pi/180;  %DOA's of signals in rad.
P=[1 1 1];         %Power of incoming signals
N=10;              %Number of array elements
K=1024;            %Number of data snapshots
d=0.5;             %Distance between elements in wavelengths
noise_var=1;       %Variance of noise
r=length(doas);  %Total number of signals

% Steering vector matrix. Columns will contain the steering vectors
% of the r signals
A=exp(-i*2*pi*d*(0:N-1)'*sin([doas(:).']));

% Signal and noise generation
sig=round(rand(r,K))*2-1; % Generate random BPSK symbols for each of the
                   % r signals
noise=sqrt(noise_var/2)*(randn(N,K)+i*randn(N,K)); %Uncorrelated noise
X=A*diag(sqrt(P))*sig+noise; %Generate data matrix

R=X*X'/K; %Spatial covariance matrix

[Q,D]=eig(R);    %Compute eigendecomposition of covariance matrix
[D,I]=sort(diag(D),1,'descend'); %Find r largest eigenvalues
Q=Q(:,I);    %Sort the eigenvectors to put signal eigenvectors first
Qs=Q(:,1:r);     %Get the signal eigenvectors
Qn=Q(:,r+1:N);   %Get the noise eigenvectors
```

```
% MUSIC algorithm

% Define angles at which MUSIC "spectrum" will be computed
angles=(-90:0.1:90);

%Compute steering vectors corresponding values in angles
a1=exp(-i*2*pi*d*(0:N-1)'*sin([angles(:).']*pi/180));
for k=1:length(angles)
  %Compute MUSIC "spectrum"
  music_spectrum(k)=(a1(:,k)'*a1(:,k))/(a1(:,k)'*Qn*Qn'*a1(:,k));
end

figure(1)
plot(angles,abs(music_spectrum))
title('MUSIC Spectrum')
xlabel('Angle in degrees')

%ESPRIT Algorithm
phi= linsolve(Qs(1:N-1,:),Qs(2:N,:));
ESPRIT_doas=asin(-angle(eig(phi))/(2*pi*d))*180/pi;

%MVDR
IR=inv(R);   %Inverse of covariance matrix
for k=1:length(angles)
  mvdr(k)=1/(a1(:,k)'*IR*a1(:,k));
end

figure(gcf+1)
plot(angles,abs(mvdr))
xlabel('Angle in degrees')
title('MVDR')

%Min norm method
alpha=Qs(1,:);
Shat=Qs(2:N,:);
ghat=-Shat*alpha'/(1-alpha*alpha');
```

```
g=[1;ghat];

for k=1:length(angles)
   minnorm_spectrum(k)=1/(abs(a1(:,k)'*g));
end

figure(gcf+1)
plot(angles,abs(minnorm_spectrum))
xlabel('Angle in degrees')
title('Min-Norm')

%Estimate DOA's using the classical beamformer
for k=1:length(angles)
   Classical(k)=(a1(:,k)'*R*a1(:,k));
end
figure(gcf+1)
plot(angles,abs(Classical))
xlabel('Angle in degrees')
title('Classical Beamformer')
```

References

[1] R. A. Monzingo and T. W. Miller, *Introduction to Adaptive Arrays*. New York: Wiley, 1980.

[2] S. Haykin, *Adaptive Filter Theory*, Englewood Cliffs, NJ: Prentice-Hall, 3rd ed., 1995.

[3] A. Alexiou and M. Haardt, "Smart antenna technologies for future wireless systems: trends and challenges," *IEEE Communications Magazine*, vol. 42, pp. 90–97, Sept. 2004. doi:10v.1109/MCOM.2004.1336725

[4] T. K. Sarkar, M.C. Wicks, M. Salazar-Palma, and R.J. Bonneau, *Smart Antenna*, Hoboken, NJ: Wiley-Interscience, 2003.

[5] J. S. Goldstein and I. S. Reed, "Reduced rank adaptive filtering," *IEEE Transactions on Signal Processing*, vol. 45, no. 2 , pp. 492–496, Feb. 1997.

[6] P. Strobach, "Fast recursive subspace adaptive ESPRIT algorithms," *IEEE* , vol. 46, pp. 2413–2430, Sept. 1998. doi:10.1109/78.709531

[7] H. Krim and M. Viberg, "Two decades of array signal processing research: the parametric approach," *IEEE Signal Processing Magazine*, vol. 13, pp. 67–94, July 1996. doi:10.1109/79.526899

[8] L. C. Godara, "Application of antenna arrays to mobile communications. Part II: Beamforming and direction of arrival considerations," *Proceedings of the IEEE*, vol. 85, pp. 1195–1245, Aug. 1997. doi:10.1109/5.622504

[9] R. Roy and T. Kailath, "ESPRIT — Estimation of signal parameters via rotational invariance techniques," *IEEE Transactions on Acoustics, Speech, Signal Processing*, vol. 37, pp. 984–995, July 1989. doi:10.1109/29.32276

[10] P. Strobach, "Fast recursive low-rank linear prediction frequency estimation algorithms," *IEEE Transactions on Signal Processing*, vol. 44, pp. 834–847, April 1996. doi:10.1109/78.492537

[11] R. O. Schmidt, "Multiple emitter location and signal parameter estimation," *IEEE Transactions on Antennas and Propagation*, vol. AP-34, pp. 276–280, Mar. 1986. doi:10.1109/TAP.1986.1143830

[12] J. Yang and M. Kaveh, "Adaptive eigensubspace algorithms for direction or frequency estimation and tracking," *IEEE Transactions on Acoustics, Speech, Signal Processing*, vol. 36, pp. 241–251, Feb. 1988. doi:10.1109/29.1516

[13] R. Badeau, G. Richard, and David, B., "Fast adaptive ESPRIT algorithm," *IEEE/SP 13th Workshop on Statistical Signal Processing*, pp. 289–294, July 2005. doi:10.1109/SSP.2005.1628608

[14] B. Yang, "Projection Approximation Subspace Tracking." *IEEE Transactions on Acoustics, Speech, Signal Processing*, vol. 45, pp. 95–107, Jan. 1995.

[15] T. S. Rappaport and J. C. Liberti Jr., *Smart Antennas for Wireless Communications: IS-95 and Third Generation CDMA Applications*, Upper Saddle River, NJ: Prentice Hall, 1999.

[16] J. Capon, "High resolution frequency-wavenumber spectral analysis," *Proceedings of the IEEE*, vol. 57, pp. 1408–1518, Aug. 1969.

[17] P. Stoica and R. Moses, *Introduction to Spectral Analysis.* Upper Saddle River, NJ: Prentice Hall, 1997.

[18] R. Kumaresan and D. W. Tufts, "Estimating the angles of arrival of multiple plane waves," *IEEE Transactions on Aerospace and Electronic Systems*, vol. AES-19, pp. 134–138, Jan. 1983.

[19] I. Ziskind and M. Wax, "Maximum likelihood localization of multiple sources by alternating projection," *IEEE Transactions on Acoustics, Speech, Signal Processing*, vol. 36, pp. 1553–1560, Oct. 1988. doi:10.1109/29.7543

[20] A. J. Barabell, "Improving the Resolution Performance of Eigenstructure-based Direction Finding Algorithms," in *Proceedings of the IEEE International Conference on Acoustics, Speech and Signal Processing,* pp. 336–339, 1983. doi:10.1109/ICASSP.1983.1172124

[21] S. Haykin, *Array Signal Processing.* Upper Saddle River, NJ: Prentice Hall, 1985.

[22] M. D. Zoltowski, M. Haardt, and C. P. Mathews, "Closed-form 2-D angle estimation with rectangular arrays in element space or beamspace via unitary ESPRIT," *IEEE Transactions Signal Processing*, vol. 44, pp. 316–328, Feb. 1996. doi:10.1109/78.485927

[23] J. Mayhan and L. Niro, "Spatial spectral estimation using multiple beam antennas," *IEEE Transactions on Antennas and Propagation*, vol. 35, pp. 897–906, Aug. 1987. doi:10.1109/TAP.1987.1144192

[24] M. D. Zoltowski, G. M. Kautz, and S. D. Silverstein, "Beamspace ROOT-MUSIC," *IEEE Transactions on Signal Processing*, vol. 41, no. 1, pp. 344–364, Feb. 1993. doi:10.1109/TSP.1993.193151

[25] G. Xu, S. D. Silverstein, R. H. Roy, and T. Kailath, "Beamspace ESPRIT," *IEEE Transactions on Signal Processing*, vol. 42, no. 2, pp. 349–356, Feb. 1994.

[26] I. Karasalo, "Estimating the covariance matrix by signal subspace averaging," *IEEE Transactions on Acoustics, Speech and Signal Processing*, vol. ASSP-34, pp. 8–12, Feb. 1986. doi:10.1109/TASSP.1986.1164779

[27] P. Strobach, "Bi-iteration SVD subspace tracking algorithms," *IEEE Transactions on Signal Processing*, vol. 45, pp. 1222–1240, May 1997. doi:10.1109/78.575696

[28] E. M. Dowling, L. P. Ammann, and R. D. DeGroat, "A TQR-iteration based adaptive SVD for real time angle and frequency tracking," *IEEE Transactions on Signal Processing*, vol. 43, pp. 914–926, Apr. 1994. doi:10.1109/78.285654

[29] V. Pisarenko, "The retrieval of harmonics from a covariance function," *Geophysical Journal of the Royal Astronomical Society*, pp. 347–366, 1973. doi:10.1111/j.1365-246X.1973.tb03424.x

[30] M. Viberg and B. Ottersten, "Sensor array processing based on subspace fitting," *IEEE Transactions on Signal Processing*, vol. 39, pp. 1110–1121, May 1991.

[31] M. Haardt and J. A. Nossek, "Unitary ESPRIT: how to obtain increased estimation accuracy with a reduced computational burden," *IEEE Transactions on Signal Processing*, vol. 43, pp. 1232–1242, May 1995. doi:10.1109/78.382406

[32] A. L. Swindlehurst, B. Ottersten, G. Xu, R. Roy, and T. Kailath, "Multiple invariance ESPRIT," *IEEE Transactions on Signal Processing*, vol. 40, pp. 867–881, Apr. 1992. doi:10.1109/78.127959

[33] A. S. Spanias, "Speech coding: A tutorial review," *Proceedings of the IEEE*, vol. 82, no. 10 pp. 1541–1582, October, 1994. doi:10.1109/5.326413

[34] A. Spanias, T. Painter, and V. Atti, *Audio Signal Processing and Coding*, NJ: Wiley-Interscience, 2006.

Additional References

P. Ioannides and C. A. Balanis, "Uniform circular arrays for smart antennas," *IEEE Antennas and Propagation Magazine*, vol. 47, pp. 192–206, Aug. 2005.

Z. Huang and C. A Balanis, "Adaptive beamforming using spherical array," *IEEE Antennas and Propagation Society International Symposium*, vol. 4, pp. 126–129, July 2005.

L. C. Godara, *Smart Antennas*, CRC Press: Boca Raton, FL, 2004.

K. Thomas, et al., "When will smart antennas be ready for the market? Part I," *IEEE Signal Processing Magazine*, vol. 87, pp. 87–92, Mar. 2005.

F. Gross, *Smart Antennas for Wireless Communications*. New York, NY: McGraw-Hill, 2005.

www.arraycomm.com [Last accessed: 19 May 2008]

K. Thomas, ed., "When will smart antennas be ready for the market? Part II — Results," *IEEE Signal Processing Magazine*, vol. 87, pp. 174–176, Nov. 2005.

G. V. Tsoulos, "Smart antennas for mobile communication systems: benefits and challenges," *Electronics & Communication Engineering Journal*, vol. 11, pp. 84–94, April 1999.

Siemens, "Advanced closed loop Tx diversity concept (eigenbeamformer)," 3GPP TSG RAN WG 1, TSGR1#14(00)0853, July 2000.

M. Chryssomallis, "Smart antennas," *IEEE Antennas and Propagation Magazine*, vol. 42, pp. 129–136, June 2000. doi:10.1109/74.848965

D. Boppana and A. Batada, "How to create beam-forming smart antennas using FPGAS," www.embedded.com, Feb. 2005.

B. D. V. Veen and K. M. Buckley, "Beamforming: a versatile approach to spatial filtering," *IEEE ASSP Magazine*, pp. 4–24, Apr. 1988.

P. Strobach, "Low-rank adaptive filters," *IEEE Transactions on Signal Processing*, vol. 44, pp. 2932–2947, Dec. 1996.

Y. Hara, C. Fujita, and Y. Kamio, "Initial weight computation method with rapid convergence for adaptive array antennas," *IEEE Transactions on Wireless Communications*, vol. 3, pp. 1902–1905, Nov. 2004. doi:10.1109/TWC.2004.837398

J. Foutz, A. Spanias, S. Bellofiore, and C. Balanis, "Adaptive eigen-projection beamforming algorithms for 1-D and 2-D antenna arrays," *IEEE Antennas and Wireless Propagation Letters*, vol. 2, pp. 62–65, 2003. doi:10.1109/LAWP.2003.811322

J. Foutz and A. S. Spanias, "Adaptive eigen-projection algorithms for 1-D and 2-D antenna arrays," *IEEE International Symposium on Circuits and Systems(ISCAS)*, vol. 2, pp. 201–204, May 2002. doi:10.1109/ISCAS.2002.1010959

J. Foutz and A. Spanias, "Adaptive modeling and control of smart antenna arrays," *20th International Conference on Modeling, Identification and Control (MIC 2001)*, Innsbruck, Austria, Feb. 2001.

S. Bellofiore, J. Foutz, R. Govindarajula, I. Bahceci, C. A. Balanis, A. S. Spanias, J. M. Capone, and T. M. Duman, "Smart antenna system analysis, integration and performance for mobile ad-hoc networks (MANETs)," *IEEE Transactions on Antennas and Propagation*, vol. 50, pp. 571–581, May 2002. doi:10.1109/TAP.2002.1011222

S. Bellofiore, C. A. Balanis, J. A. Foutz, and A. S. Spanias, "Smart antenna system introduction, integration and performance for mobile communication networks. Part I: Overview and antenna design," *IEEE Antennas and Propagation Magazine*, vol. 44, pp.145–154, June 2002.

S. Bellofiore, C. A. Balanis, J. A. Foutz, and A. S. Spanias, "Smart antenna for mobile communication networks. Part II: Beamforming and network throughput," *IEEE Antennas and Propagation Magazine*, vol. 44, pp. 106–114, Aug. 2002.

A. J. Van Der Veen, E. D. Deprettere, and A. L. Swindlehurst, "Subspace-based signal analysis using singular value decomposition," *Proceedings of the IEEE*, vol. 81, no. 9, pp. 1277–1308, Sep. 1993. doi:10.1109/5.237536

L. L. Scharf, "The SVD and reduced rank signal processing," *Signal Processing*, vol. 25, no. 2, pp. 113–133, Nov. 1991. doi:10.1016/0165-1684(91)90058-Q

L. L. Scharf and B. D. Van Veen, "Low rank detectors for Gaussian random vectors," *IEEE Transactions on Acoustics, Speech and Signal Processing*, vol. ASSP-35, no. 11, pp.1579–1582, Nov. 1987. doi:10.1109/TASSP.1987.1165076

J. Goldstein, L. Reed, and L. Scharf, "A multistage representation of the Wiener filter based on orthogonal projections," *IEEE Transactions on Information Theory*, vol. 44, pp. 2943–2959, Nov. 1998.

X. Wang and H. V. Poor, "Blind multiuser detection: a subspace approach," *IEEE Transactions on Information Theory*, vol. 44, pp. 677–690, Mar. 1998.

M. Honig and M. Tsatsanis, "Adaptive techniques for multiuser CDMA receivers," *IEEE Signal Processing Magazine*, vol. 17, no. 3, pp. 49–61, May 2000. doi:10.1109/79.841725

S. Kraut, L. L. Scharf, and T. McWhorter, "Adaptive subspace detectors," vol. 49, pp. 1–16, Jan. 2001. doi:10.1109/78.890324

X. Wang and H. V. Poor, "Blind multiuser detection: a subspace approach," *IEEE Transactions on Information Theory*, vol. 44, pp. 677–690, Mar. 1998.

G. Wu, H. Wang, M. Chen, and S. Cheng, "Performance comparison of space-time spreading and space-time transmit diversity in CDMA2000," *IEEE Vehicular Technology Conference*, vol. 1, pp. 442–446, 2001.

V. Varadarajan and J. Krolik, "Array shape estimation tracking using active sonar reverberation," *IEEE Transactions on Aerospace and Electronic Systems*, vol. 40, pp. 1073–1086, July 2004. doi:10.1109/TAES.2004.1337475

J. L. Yu and C. C. Yeh, "Generalized eigenspace-based beamformers," *IEEE Transactions on Signal Processing*, vol. 43, pp. 2453–2461, Nov. 1995.

D. Segovia-Vargas, F. Inigo, and M. Sierra-Perez, "Generalized eigenspace beamformer based on CG-Lanczos algorithm," *IEEE Transactions on Antennas and Propagation*, vol. 51, pp. 2146–2154, Aug. 2003. doi:10.1109/TAP.2003.814744

A. Chang, C. Chiang, and Y. Chen, "A generalized eigenspace-based beamformer with robust capabilities," *IEEE International Conference Phased Array Systems and Technology*, pp. 553–556, May 2000.

W. S. Youn and C. K. Un, "Robust adaptive beamforming based on the eigenstructure method," *IEEE Transactions on Signal Processing*, vol. 42, pp. 1543–1547, June 1994.

W. S. Youn and C. K. Un, "Eigenstructure method for robust array processing," *Electronics Letters*, vol. 26, pp. 678–680, May 1990. doi:10.1049/el:19900444

Y. Zhao, Z. Wang, and S. Zhang, "An eigenspace-based algorithm for adaptive antenna sidelobe cancellation," *International Conference on Signal Processing*, vol. 3, pp. 2086–2089, Sept. 2004.

Z. Yongbo and Z. Shouhong, "A modified eigenspace-based algorithm for adaptive beamforming," *WCCC-ICSP*, vol. 1, pp. 468–471, Aug. 2000.

J. T. Reagen and T. Ogunfunmi, "A LORAF-based coherent sidelobe canceller for narrowband adaptive beamforming applications," *Asilomar Conference on Signals, Systems and Computers*, vol. 1, pp. 556–560, Nov. 1996. doi:10.1109/ACSSC.1996.601082

A. M Haimovich and Y. Bar-Ness, "An eigenanalysis interference canceler," *IEEE Transactions on Signal Processing*, vol. 39, pp. 76–84 , Jan. 1991. doi:10.1109/78.80767

Y. Zhao and S. Zhang, "An interference canceler robust to beam pointing error," *Sensor Array and Multichannel Signal Processing Workshop*, pp. 48–52, Aug. 2002.

S. J. Yu and J. H. Lee, "The statistical performance of eigenspace-based adaptive array beamformers," *IEEE Transactions on Antennas and Propagation*, vol. 44, pp. 665–671, May 1996.

L. Chang and C. C. Yeh, "Performance of DMI and eigenspace-based beamformers," *IEEE Transactions on Antennas and Propagation*, vol. 40, pp. 1336–1347, Nov. 1992. doi:10.1109/8.202711

B. D. Van Veen, "Eigenstructure based partially adaptive array design," *IEEE Transactions on Antennas and Propagation*, vol. 36, pp. 357–362, March 1988. doi:10.1109/8.192118

T. McWhorter, "Fast rank-adaptive beamforming," *IEEE Sensor Array and Multichannel Signal Processing Workshop*, pp. 63–67, March 2000. doi:10.1109/SAM.2000.877969

J. H. Lee and C. C. Lee, "Analysis of the performance and sensitivity of an eigenspace-based interference canceler," *IEEE Transactions on Antennas and Propagation*, vol. 48, pp. 826–835, May 2000.

L. Chang, "A real-valued adaptive eigencomponents updated algorithm for beamforming," *ICCS*, vol. 3, pp. 897–901, Nov. 1994. doi:10.1109/ICCS.1994.474275

W. C. Lee and C. Seugwon, "Adaptive beamforming algorithm based on eigen-space method for smart antennas, *IEEE Communications Letters*, vol. 9, pp. 888–890, Oct. 2005.

C. C. Lee and J. H. Lee, "Eigenspace-based adaptive array beamforming with robust capabilities," *IEEE Transactions on Antennas and Propagation*, vol. 45, pp. 1711–1716, Dec. 1997.

A. K. Sadek, W. Su, and K. J. R. Liu, "Eigen-selection approach for joint beamforming and space-frequency coding in MIMO-OFDM systems with spatial correlation feedback," *IEEE Workshop on Signal Processing Advances in Wireless Communications*, pp. 565–569, June 2005. doi:10.1109/SPAWC.2005.1506203

N. G. Nair and A. S. Spanias, "Gradient eigenspace projections for adaptive filtering," *Midwest Symposium on Circuits and Systems*, vol. 1, pp. 259–263, Aug. 1995.

N. G. Nair and A. S. Spanias, "Fast adaptive algorithms using eigenspace projections," *Asilomar Conference on Signals, Systems and Computers*, vol. 2, pp. 1520–1524, Nov. 1994. doi:10.1109/ACSSC.1994.471712

Y. Hua, T. Chen, and Y. Miao, "A unifying view of a class of subspace tracking methods," in *Proceedings of the Symposium on Image, Speech, Signal Processing, and Robotics*, vol. 2, pp. 27–32, Sept. 1998.

E. Oja, "Neural networks, principal components, and subspaces," *International Journal of Neural Systems*, vol. 1, pp. 61–68, 1989.

E. Oja, "A simplified neuron model as a principal component analyzer," *Journal of Mathematical Biology*, vol. 15, pp. 267–273, 1982.

S. C. Douglas, "Numerically-robust adaptive subspace tracking using Householder transformations," *IEEE Sensor Array and Multichannel Signal Processing Workshop*, pp. 499–503, Mar. 2000. doi:10.1109/SAM.2000.878059

S. C. Douglas and X. Sun, "Designing orthonormal subspace tracking algorithms," *Asilomar Conference on Signals, Systems and Computers*, vol. 2, pp. 1441–1445, Nov. 2000. doi:10.1109/ACSSC.2000.911229

M. Wax and T. Kailath, "Detection of signals by information theoretic criteria," *IEEE Transactions on Acoustics Speech, Signal Processing*, vol. ASSP-33, pp. 387–392, Apr. 1985. doi:10.1109/TASSP.1985.1164557

B. Yang, "An extension of the PASTd algorithm to both rank and subspace tracking," *IEEE Signal Processing Letters*, vol. 2, pp. 179–182, Sept. 1995.

K. B. Yu, "Recursive updating the eigenvalue decomposition of a covariance matrix," *IEEE Transactions on Signal Processing*, vol. 39, pp. 1136–1145, May 1991. doi:10.1109/78.80968

R. Schreiber, "Implementation of adaptive array algorithms," *IEEE Transactions on Acoustics, Speech, Signal Processing*, vol. ASSP-34, pp. 1038–1045, Oct. 1986. doi:10.1109/TASSP.1986.1164943

E. C. Real, D. W. Tufts, and J. W. Cooley, "Two algorithms for fast approximate subspace tracking," *IEEE Transactions on Signal Processing*, vol. 47, no. 7, pp. 1936–1945, July 1999. doi:10.1109/78.771042

D. B. Badeau and G. Richard, "Fast approximated power iteration subspace tracking," *IEEE Transactions on Signal Processing*, vol. 53, pp. 2931–2941, Aug. 2005.

B. Champagne and Q. Liu, "Plane rotation-based EVD updating schemes for efficient subspace tracking," *IEEE Transactions on Signal Processing*, vol. 46, pp. 1886–1900, July 1998.

Z. Fu, E. M. Dowling, and R. D. Degroat, "Systolic arrays for spherical subspace tracking," *Conference Record of The Twenty-Seventh Asilomar Conference on Signals, Systems and Computers*, vol. 1, pp. 766–770, Nov. 1993.

F. Vanpoucke and M. Moonen, "Parallel and stable spherical subspace tracking," in *Proceedings of the IEEE International Conference on Acoustics, Speech, Signal Processing*, vol. 3, pp. 2064–2067, May 1995. doi:10.1109/ICASSP.1995.478480

B. Yang, "A systolic architecture for gradient based adaptive subspace tracking algorithms," *Workshop on VLSI Signal Processing*, pp. 516–524, Oct. 1993.

F. Xu and A. N. Willson Jr., "Novel systolic architectures for signal subspace tracking," *IEEE Midwest Symposium on Circuits and Systems*, pp. 880–883, vol. 2, Aug. 2000.

F. Xu and A. N. Willson Jr., "Local stability analysis and systolic implementation of a subspace tracking algorithm," in *Proceedings of the IEEE International Conference on Acoustics, Speech, Signal Processing*, pp. 3881–3884, vol. 6, May 2001.

F. Xu and A. N. Willson, "Efficient hardware architectures for eigenvector and signal subspace estimation," *IEEE Transactions on Circuits and Systems*, vol. 51, pp. 517–525, March 2004.

D. Rabideau and A. Steinhardt, "Fast subspace tracking using coarse grain and fine grain parallelism," in *Proceedings of the IEEE International Conference on Acoustics, Speech, Signal Processing*, vol. 5, pp. 3211–3214, May 1995.

T. Gustafsson and M. Viberg, "Instrumental variable subspace tracking with applications to sensor array processing and frequency estimation," *8th IEEE Signal Processing Workshop on Statistical Signal and Array Processing*, pp. 78–81, June 1996. doi:10.1109/SSAP.1996.534824

T. Gustafsson, "Instrumental variable subspace tracking using projection approximation," *IEEE Transactions on Signal Processing*, vol. 46, pp. 669–681, March 1998. doi:10.1109/78.661334

B. Yang, "Convergence analysis of the subspace tracking algorithms PAST and PASTd," in *Proceedings of the IEEE International Conference on Acoustics, Speech, Signal Processing*, vol. 3, pp. 1759–1762, May 1996.

J. Lee and B. Kyung-Bin, "Numerically stable fast sequential calculation for projection approximation subspace tracking," *IEEE International Symposium on Circuits and Systems*, vol. 3, pp. 484–487, June 1999. doi:10.1109/ISCAS.1999.778888

L. Jun-Seek, S. Seongwook, and S. Keeng-Mog, "Variable forgetting factor PASTd algorithm for time-varying subspace estimation," *Electronics Letters*, vol. 36, pp. 1434–1435, Aug. 2000.

Y. Wen, S. C. Chan, and K. L. Ho, "Robust subspace tracking in impulsive noise," *IEEE International Conference on Communications*, vol. 3, pp. 892–896, June 2001. doi:10.1109/ICC.2001.937366

S. C. Chan, Y. Wen, and K. L. Ho, "A robust past algorithm for subspace tracking in impulsive noise," *IEEE Transactions on Signal Processing*, vol. 54, pp. 105–116, Jan. 2006.

Y. Wen, S. C. Chan, and K. L. Ho, "A robust subspace tracking algorithm for subspace-based blind multiuser detection in impulsive noise," *International Conference on Digital Signal Processing*, vol. 2, pp. 1289–1292, July 2002. doi:10.1109/ICDSP.2002.1028329

K. Berberidis, "Block subspace updating algorithms for tracking directions of coherent signals in SDMA mobile systems," in *Proceedings of the International Conference on Electronics, Circuits and Systems*, vol. 3, pp. 1091–1094, Sept. 2002. doi:10.1109/ICECS.2002.1046441

S. Buzzi, M. Lops, and A. Pauciullo, "Iterative cyclic subspace tracking for blind adaptive multiuser detection in multirate CDMA systems," *IEEE Transactions on Vehicular Technology*, vol. 52, pp. 1463–1475, Nov. 2003. doi:10.1109/TVT.2003.816637

H. Zhang, G. Ren, H. Zhang, and J. Zhang, "An improved OPAST algorithm for spatio-temporal multiuser detection technique based on subspace tracking," *The Ninth International Conference on Communications Systems*, pp. 401–404, Sept. 2004.

S. Bartelmaos, K. Abed-Meraim, and S. Attallah, "Mobile localization using subspace tracking Communications," *Asia-Pacific Conference*, pp. 1009–1013, Oct. 2005. doi:10.1109/APCC.2005.1554216

L. Lin, "Subspace based blind multiuser detection asynchronous MC-CDMA systems," in *Proceedings of the International Conference on Wireless Communications, Networking and Mobile Computing*, vol. 1, pp. 639–643, Sept. 2005.

G. Xu and T. Kailath, "Fast subspace decomposition," *IEEE Transactions on Signal Processing*, vol. 42, pp. 539–551, March 1994.

G. H. Golub and C. F. Van Loan, *Matrix Computations*, Baltimore, MD: John Hopkins Press, 1996.

F. Zuqiang and E. M. Dowling, "Conjugate gradient projection subspace tracking," *IEEE Transactions on Signal Processing*, vol. 45, no. 6, pp. 1664–1668, June 1997. doi:10.1109/78.600010

Y. Miao and Y. Hua, "Fast subspace tracking by a novel information criterion," *Conference Record of the Thirty-First Asilomar Conference on Signals, Systems & Computers*, vol. 2, pp. 1312–1316, Nov. 1997.

Y. Miao and Y. Hua, "Fast subspace tracking and neural network learning by a novel information criterion," *IEEE Transactions on Signal Processing*, vol. 46, no. 7, pp. 1967–1979, July 1998. doi:10.1109/78.700968

W. C. Lee, S. T. Park Il, W. Cha, and D. H. Youn, "Adaptive spatial domain forward-backward predictors for bearing estimation," IEEE Transactions on Acoustics, Speech, Signal Processing, vol. 38, pp. 1105–1109, July 1990.

R. Govindarajula, and J. M. Capone, "Enhancing the capacity of ad-hoc networks with smart antennas," Presentation to Smart Antennas for Future Reconfigurable Wireless Communication Networks, Oct. 2000, NSF Grant No. ECS-9979403.

D. E. Dudgeon and R. M. Mersereau, Multidimensional Digital Signal Processing, Englewood Cliffs, NJ: Prentice Hall, 1984.

J. W. Demmel, Applied Numerical Linear Algebra, SIAM, Philadelphia, 1997.

M. Viberg and B. Ottersten, "Sensor array processing based on subspace fitting," IEEE Transactions on Signal Processing, vol. 39, pp. 1110–1121, May 1991.

R. D. DeGroat, "Noniterative subspace tracking," IEEE Transactions on Signal Processing, vol. 40, pp. 571–577, Mar. 1992.

D. R. Fuhrmann, "Rotational search methods for adaptive Pisarenko harmonic retrieval," IEEE Transactions on Signal Processing, ASSP-34, pp. 1550–1565, Dec. 1986.

N. L. Owsley, "Adaptive data orthogonalization," in Proceedings of the IEEE International Conference on Acoustics, Speech, Signal Processing, pp. 109–112, April 1978.

P. Comon and G. H. Golub, "Tracking a few extreme singular values and vectors in signal processing," Proceedings of the IEEE, vol. 78, pp. 1327–1343, Aug. 1990.

Y. Hua, Y. Xiang, T. Chen, K. Abed-Meraim, and Y. Miao, "Natural power method for fast subspace tracking," IEEE Signal Processing Soc. Workshop, Neural Networks for Signal Processing IX, pp. 176–185, 1999.

J. R. Bunch, C. P. Nielsen, and D. C. Sorensen, "Rank one modification of the symmetric eigenproblem," Numerische Mathematik, vol. 3, pp. 111–129, 1978.

H. Chen, T. K. Sarker, S. A. Dianat, and J. D. Brule, "Adaptive spectral estimation by the conjugate gradient method," IEEE Transactions on Acoustics, Speech, Signal Processing, vol. ASSP-34, pp. 272–284, Apr. 1986.

Z. Fu and M. Dowling, "Conjugate gradient eigenstructure tracking for adaptive spectral estimation," IEEE Transactions on Signal Processing, vol. 43, pp. 1151–1160, May 1995.

L. Scharf, Statistical Signal Processing, Reading, MA: Addison Wesley, 1990.

S. L. Marple, Digital Spectral Analysis with Applications, Upper Saddle River, NJ: Prentice Hall, 1987.

List of Symbols

$\mathbf{x}_n$	$N \times 1$ data vector at time index n
n	Discrete time index
N	Dimension of data vector $\mathbf{x}_n$
r	Number of signals present in the linear data model of $\mathbf{x}_n$
$\mathbf{R}_{xx}$	Autocorrelation matrix associated with $\mathbf{x}_n$
$\mathbf{R}_{xx}(n)$	Estimate of $\mathbf{R}_{xx}$ at time n
$\mathbf{I}_{n \times k}$	The $n \times k$ identity matrix
$\mathbf{I}_n$	The $n \times n$ identity matrix
$\mathbf{A}$	$N \times r$ matrix whose columns are steering vectors of incoming signals
σ^2	Variance of the white, Gaussian noise
$\mathbf{QDQ}^{\mathrm{H}}$	The eigendecomposition of $\mathbf{R}_{xx}$, where $\mathbf{QQ}^{\mathrm{H}} = \mathbf{I}$, and $\mathbf{Q}$ can be partitioned as $[\mathbf{Q}_s \ \mathbf{Q}_n]$, where $\mathbf{Q}_s$ is $N \times r$ and $\mathbf{Q}_n$ is $N \times (N - r)$. $\mathbf{D} = \mathrm{diag}\{\sigma_0^2, \sigma_1^2, \ldots, \sigma_{N-1}^2\}$, where $\sigma_0^2 \geq \sigma_1^2 \geq \ldots \geq \sigma_{r-1}^2 > \sigma_r^2 = \sigma_{r+1}^2 = \ldots = \sigma_{N-1}^2$
$\mathbf{Q}_s(n)$	$\mathbf{Q}_s(n)$ is the estimate of $\mathbf{Q}_s$ at time index n
$\mathbf{X}(n)$	$n \times N$ data matrix whose ith row is $\mathbf{x}_i^{\mathrm{H}}$
$\mathbf{USV}^{\mathrm{H}}$	Singular value decomposition of the data matrix $\mathbf{X}(n)$, with $\mathbf{S} = \mathrm{diag}\{\sigma_0, \sigma_1, \ldots, \sigma_{N-1}\}$. The columns of $\mathbf{U}$ are the left singular vectors of $\mathbf{X}$ and the columns of $\mathbf{V}$ are the right singular vectors. The matrix $\mathbf{V}$ can be partitioned as $\mathbf{V} = [\mathbf{V}_s \ \mathbf{V}_n]$ with $\mathbf{V}_s = \mathbf{Q}_s$ and $\mathbf{V}_n = \mathbf{Q}_n$
$\mathbf{z}_n$	$\mathbf{z}_n = \mathbf{Q}_s(n)\mathbf{x}_n$, $r \times 1$ compressed data vector
λ	Wavelength of bandpass signal
c	Speed of light
D	Distance between elements of uniform linear array in meters
d	Distance between elements of a uniform linear array in wavelengths
$\mathbf{w}_n$	$N \times 1$ vector of complex beamformer weights

List of Acronyms

DOA	Direction of arrival
ESPRIT	Estimation of signal parameters via rotational invariance techniques
EVD	Eigenvalue decomposition
MIL	Matrix inversion lemma
MUSIC	Multiple signal classification
SVD	Singular value decomposition
SW	Sliding window
ULA	Uniform linear array
URA	Uniform rectangular array
DFT	Discrete Fourier transform

Author Biography

Jeff Foutz received his B.S., M.S., and Ph.D. degrees in electrical engineering from Arizona State University in 1998, 2001, and 2007, respectively. At Arizona State, he performed research in the field of adaptive arrays, including direction of arrival estimation, beamforming, and subspace tracking. From 2002 to 2005, he was with Motorola/Freescale Semiconductor where he worked on video processing for cable television and telematics applications. Since 2007, he has been with GE Healthcare working as a medical image processing engineer.

Andreas Spanias is a professor in the Department of Electrical Engineering, Fulton School of Engineering at Arizona State University. He is also the director of the SenSIP consortium. His research interests are in the areas of adaptive signal processing, speech processing, and audio sensing. Prof. Spanias has collaborated with Intel Corporation, Sandia National Labs, Motorola, Texas Instruments, DTC, Freescale, Microchip, and Active Noise and Vibration Technologies. He and his student team developed the computer simulation software Java-DSP (J-DSP; ISBN 0-9724984-0-0). He is author of two textbooks, *Audio Processing and Coding* by Wiley and *DSP: An Interactive Approach*. He received the 2003 Teaching Award from the IEEE Phoenix section for the development of J-DSP. He has served as associate editor of the *IEEE Transactions on Signal Processing* and as *General Cochair* of the 1999 International Conference on Acoustics Speech and Signal Processing (ICASSP-99) in Phoenix. He also served as the IEEE Signal Processing Vice President for Conferences and is currently member-at-large of the IEEE SPS Conference Board. Prof. Spanias is co-recipient of the 2002 IEEE Donald G. Fink paper prize award and was elected fellow of the IEEE in 2003. He served as distinguished lecturer for the IEEE Signal Processing Society. He is currently the editor for the Morgan & Claypool Publishers series on DSP algorithms and software.

Mahesh K. Banavar is a graduate student at Arizona State University. He received his B.E. degree in telecommunications engineering from Visvesvaraya Technological University, Karnataka, India, in 2005 and his M.S. degree in electrical engineering from Arizona State University in 2008. He is currently a Ph.D. student at Arizona State University specializing in signal processing and communications and doing research in wireless communications and sensor networks. He is a member of the Eta Kappa Nu honor society and a student member of the IEEE.

Printed in the United States
by Baker & Taylor Publisher Services

Synthesis Lectures on Antennas

Jeffrey Foutz · Andreas Spanias · Mahesh K. Banavar

Narrowband Direction of Arrival Estimation for Antenna Arrays

This book provides an introduction to narrowband array signal processing, classical and subspace-based direction of arrival (DOA) estimation with an extensive discussion on adaptive direction of arrival algorithms. The book begins with a presentation of the basic theory, equations, and data models of narrowband arrays. It then discusses basic beamforming methods and describes how they relate to DOA estimation. Several of the most common classical and subspace-based direction of arrival methods are discussed. The book concludes with an introduction to subspace tracking and shows how subspace tracking algorithms can be used to form an adaptive DOA estimator. Simulation software and additional bibliography are given at the end of the book.

ISBN 978-3-031-00409-4

▶ springer.com